D0710520

ELEMENTS OF
ENVIRONMENTAL
CHEMISTRY

THE WILEY BICENTENNIAL—KNOWLEDGE FOR GENERATIONS

*E*ach generation has its unique needs and aspirations. When Charles Wiley first opened his small printing shop in lower Manhattan in 1807, it was a generation of boundless potential searching for an identity. And we were there, helping to define a new American literary tradition. Over half a century later, in the midst of the Second Industrial Revolution, it was a generation focused on building the future. Once again, we were there, supplying the critical scientific, technical, and engineering knowledge that helped frame the world. Throughout the 20th Century, and into the new millennium, nations began to reach out beyond their own borders and a new international community was born. Wiley was there, expanding its operations around the world to enable a global exchange of ideas, opinions, and know-how.

For 200 years, Wiley has been an integral part of each generation's journey, enabling the flow of information and understanding necessary to meet their needs and fulfill their aspirations. Today, bold new technologies are changing the way we live and learn. Wiley will be there, providing you the must-have knowledge you need to imagine new worlds, new possibilities, and new opportunities.

Generations come and go, but you can always count on Wiley to provide you the knowledge you need, when and where you need it!

WILLIAM J. PESCE
PRESIDENT AND CHIEF EXECUTIVE OFFICER

PETER BOOTH WILEY
CHAIRMAN OF THE BOARD

ELEMENTS OF ENVIRONMENTAL CHEMISTRY

Ronald A. Hites

Indiana University

BICENTENNIAL

1807

WILEY

2007

BICENTENNIAL

WILEY-INTERSCIENCE
A JOHN WILEY & SONS, INC., PUBLICATION

Published by John Wiley & Sons, Inc., Hoboken, New Jersey
Published simultaneously in Canada

Limit of Liability/Disclaimer of Warranty: While the publisher and author have used their best efforts in preparing this book, they make no representations or warranties with respect to the accuracy or completeness of the contents of this book and specifically disclaim any implied warranties of merchantability or fitness for a particular purpose. No warranty may be created or extended by sales representatives or written sales materials. The advice and strategies contained herein may not be suitable for your situation. You should consult with a professional where appropriate. Neither the publisher nor author shall be liable for any loss of profit or any other commerical damages, including but not limited to special, incidental, consequential, or other damages.

For general information on our other products and services or for technical support, please contact our Customer Care Department within the United States at (800) 762-2974, outside the United States at (317) 572-3993 or fax (317) 572-4002.

Wiley also publishes its books in a variety of electronic formats. Some content that appears in print may not be available in electronic formats. For more information about Wiley products, visit our web site at www.wiley.com.

Wiley Bicentennial Logo: Richard J. Pacifico

Library of Congress Cataloging-in-Publication Data:

Hites, R. A.
    Elements of environmental chemistry/Ronald A. Hites.
        p. cm.
    Includes index.
    ISBN 978-0-471-99815-0 (cloth)
1. Environmental chemistry. I. Title.
TD193.H58 2007
577′.14–dc22   2006038732

Printed in the United States of America

10 9 8 7 6 5 4 3 2

To my family
Bonnie
Veronica, Karin, and David

A Note on the Cover

The illustrations on the cover represent the four "elements" in an environmental chemist's periodic table: air, earth, fire, and water. This bit of whimsy was suggested by a Sidney Harris cartoon appearing in his book *What's So Funny About Science?* (Wm. Kaufmann, Inc., Los Altos, CA, 1977).

CONTENTS

PREFACE

Many chemistry and environmental science departments now feature a course on environmental chemistry, and several textbooks support these courses. As you might expect, the coverage and quality of these textbooks varies—in some cases dramatically. Although it is obviously a matter of opinion (depending on the instructor's background and skills), it seems to me that a good textbook on environmental chemistry should include, at a minimum, the following topics: steady- and non-steady-state modeling, chemical kinetics, stratospheric ozone, photochemical smog, the greenhouse effect, carbonate equilibria, the application of partition coefficients (K_{ow}, etc.), pesticides, and toxic metals. In addition, we must always remember that environmental chemistry is a quantitative science; thus, a good textbook for environmental chemistry should also develop students' quantitative skills by providing numerous real-world problems.

This book aims for a quantitative approach to most of these topics. In fact, one could think of this book as providing the student with the essence of environmental chemistry *and* with a toolbox for solving problems. The latter skills are transferable to other fields beyond environmental chemistry. Hopefully, this book will allow students to understand methods of problem solving in the context of environmental chemistry and provide the basic concepts of environmental chemistry, so that these problem-solving techniques can be used to understand even complex environmental challenges.

This is a short book, in some ways modeled after *Elements of Style* by W. Strunk and E.B. White and *Consider a Spherical Cow* by J. Harte. Like those classic texts, the goal of this textbook is to be tutorial and informal;[1] thus, the text features many quantitative story problems (indicated by bold font). For each problem, a strategy is developed and the solution provided. This book is not intended to be read as a novel. It is an interactive textbook, and it is intended to be read with a pencil in hand so that the student can follow along with the problem statement, the strategy for solving the problem and the calculations used in arriving at an answer. "Reading" this book will do the student little good without actually doing the problems. It is not sufficient to say to yourself, "I could do that problem if I really had to." You must do all of the problems if you are going to learn this material. In addition to the problems in the text, each chapter ends with a problem set covering the quantitative aspects of the material. Answers to these problems are at the back of the book, and full solutions to the problem set questions are available on the John Wiley & Sons web site at http://www.wiley.com. Do all the problems in all the problem sets!

This book stems from a course I taught for 25 years at Indiana University to first-year graduate students, who for the most part came into our program with under-

---

[1] Note the occasional "jokes" in the footnotes.

graduate degrees in biology. Thus, as a stand-alone text, this book is suitable for a one-semester course (particularly if supplemented with a few lectures on the instructor's favorite environmental topics) aimed at upper-level undergraduate chemistry or chemical engineering majors or at first-year graduate students with only a modest physical science background. This book would also make a good companion text for courses wanting to add a patina of environmental topics to an otherwise dull subject and for courses that cover other environmental sciences such as ecology. Because of its tutorial nature, it would also make a good self-study text for entry-level professionals. A little calculus will help the reader follow the exposition in a few places, but it is not really necessary.

I thank Philip S. Stevens and Jeffrey R. White for their insightful comments on parts of the text. I also thank the hundreds of students who used this material in my classes over the years and who were not shy in explaining to me where the material was deficient. Nevertheless, errors likely remain, and I take full responsibility for them.

I would be happy to hear from you. If I have omitted your favorite topic or if I have been singularly unclear about some topic, please let me know. If you disagree with my problem set solutions, please let me know.

Ronald A. Hites

Bloomington, Indiana
September 2006
HitesR@Indiana.edu

.

CHAPTER 1

SIMPLE TOOL SKILLS

There are a variety of little tasks that will occur over and over again as we work through quantitative problems, and we need to master them first. These tasks include unit conversions, estimating, the ideal gas law, and stoichiometry.

1.1 UNIT CONVERSIONS

There are several important prefixes you should know and should probably memorize.

Femto	(f)	10^{-15}
Pico	(p)	10^{-12}
Nano	(n)	10^{-9}
Micro	(μ)	10^{-6}
Milli	(m)	10^{-3}
Centi	(c)	10^{-2}
Kilo	(k)	10^{3}
Mega	(M)	10^{6}
Giga	(G)	10^{9}
Tera	(T)	10^{12}

For example, a nanogram is 10^{-9} g, and a kilometer is 10^{3} m.

Elements of Environmental Chemistry, by Ronald A. Hites
Copyright © 2007 John Wiley & Sons, Inc.

For those of us forced by convention or national origin to work with the so-called English units, there are some other handy conversion factors you should know:

1 pound (lb) = 454 g
1 inch (in.) = 2.54 cm
12 in. = 1 foot (ft)
1 m = 3.28 ft
1 mile = 5280 ft = 1609 m
3.79 L = 1 U.S. gallon (gal), liquids only

There are some other common conversion factors that link length units to more common volume and area units:

$$1 \text{ m}^3 = 10^3 \text{ L}$$
$$1 \text{ km}^2 = (10^3 \text{ m})^2 = 10^6 \text{ m}^2 = 10^{10} \text{ cm}^2$$

One more unit conversion that we will find very helpful is

$$1 \text{ tonne (t)} = 10^3 \text{ kg} = 10^6 \text{ g}$$

Yes, we will spell metric *tonne* like this to distinguish it from 1 U.S. short ton, which is 2000 lb. One short ton equals 0.91 metric tonnes.

Let us do some simple unit conversion examples. The point is to carry along the units as though they were algebra and cancel out things as you go. Always write down your unit conversions! I cannot begin to count the number of people who looked foolish at public meetings because they tried to do unit conversions in their heads.

Human head hair grows about one half of an inch per month. How much hair grows in 1 s; please use metric units?

Strategy. Let us convert inches to meters and months to seconds. Then depending on how small the result is, we can select the right length units.

$$\text{Rate} = \left(\frac{0.5\text{ in.}}{\text{month}}\right)\left(\frac{2.54\text{ cm}}{\text{in.}}\right)\left(\frac{\text{m}}{10^2\text{ cm}}\right)$$

$$\times\left(\frac{\text{month}}{31\text{ days}}\right)\left(\frac{\text{day}}{24\text{h}}\right)\left(\frac{\text{h}}{60\text{ min}}\right)\left(\frac{\text{min}}{60\text{ s}}\right)$$

$$= 4.7 \times 10^{-9}\text{m/s}$$

If scientific notation is confusing to you, learn to use it.[1] We can put this hair growth rate in more convenient units:

$$\text{Rate} = \left(\frac{4.7 \times 10^{-9}\text{m}}{\text{s}}\right)\left(\frac{10^9\text{ nm}}{\text{m}}\right) = 4.7\text{ nm/s}$$

---

[1] We will use scientific notation throughout this book because it is easier to keep track of very big or very small numbers. For example, in the calculation we just did, we would have ended up with a growth rate of 0.000,000,0047 m/s in regular notation; that number is difficult to read and prone to error in transcription (you have to count the zeros accurately). To avoid this problem, we give the number followed by 10 raised to the correct power. It is also easier to multiply and divide numbers in this format. For example, it is tricky to multiply 0.000,000,0047 by 1000,000,000, but it is easy to multiply 4.7×10^{-9} by 1×10^9 by multiplying the leading numbers $(4.7 \times 1 = 4.7)$ and by adding the exponents of 10 $(-9 + 9 = 0)$ giving a result of $4.7 \times 10^0 = 4.7$.

This is not much, but it obviously mounts up second after second.

A word on significant figures: In the above result, the input to the calculation was 0.5 in. per month, a datum with only one significant figure. Thus, the output from the calculation should not have more than one significant figure and should have been given as 5 nm/s. In general, one should use a lot of significant figures inside the calculation, but round off the answer to the correct number of figures at the end. With a few exceptions, one should be suspicious of environmental results having four or more significant figures; in most cases, two will do.

The total amount of sulfur released into the atmosphere per year by the burning of coal is about 75 million tonnes. Assuming this were all solid sulfur, how big a cube would this occupy? You need the dimension of each side of the cube in feet. Assume the density of sulfur is twice that of water.

Strategy. Ok, this is a bit more than just converting units. We have to convert weight to volume, and this requires knowing the density of sulfur; density has units of weight per unit volume, which in this case is given to be twice that of water. As you may remember, the density of water is 1 g/cm^3, so the density of sulfur is 2 g/cm^3. Once we know the volume of sulfur, we can take the cube root of that volume and get the side length of a cube holding that volume.

$$V = (7.5 \times 10^7 \, \text{t}) \left(\frac{\text{cm}^3}{2 \, \text{g}} \right) \left(\frac{10^6 \, \text{g}}{\text{t}} \right) = 3.8 \times 10^{13} \, \text{cm}^3$$

$$\text{Side} = \sqrt[3]{3.8 \times 10^{13} \text{cm}^3} = 3.35 \times 10^4 \, \text{cm} \left(\frac{\text{m}}{10^2 \, \text{cm}} \right)$$

$$= 335 \, \text{m}$$

$$\text{Side} = 335 \, \text{m} \left(\frac{3.28 \, \text{ft}}{\text{m}} \right) = 1100 \, \text{ft}$$

This is huge. It is a cube as tall as the Empire State Building on all three sides. Pollution gets scary if you think of it as being all in one place rather than diluted by the Earth's atmosphere.

1.2 ESTIMATING

We often need order of magnitude guesses for many things in the environment. This is an important skill, so let us start with a couple of examples.

How many cars are there in the United States and in the world?

Strategy. Among our friends and families, it seems like about every other person has a car. If we know the population of the United States, then we can use this 0.5 cars per person conversion factor to get the number of cars in the United States. It would be wrong to use this 0.5 cars per person for the rest of the world (e.g., there are not 500 million cars in China—yet), but we might just use a multiplier based on the size of the economy of the United States versus the world. We know that the U.S. economy is roughly one third that of the whole world; hence, we can multiply the number of cars in the United States by 3 to estimate the number of cars in the world.

In the United States, there are now about 295 million people and almost every other person has a car; thus,

$$2.95 \times 10^8 \times 0.5 = 1.5 \times 10^8 \text{ cars in the United States}$$

The U.S. economy is about one third of the world's economy; hence, the number of cars in the world is

$$3 \times 1.5 \times 10^8 \approx 5 \times 10^8$$

The real number is not known with much precision, but in 2005, it is likely on the order of $\sim 6 \times 10^8$ cars. Thus, our estimate is a bit low, but it is certainly in the right ballpark. Of course, this number will increase dramatically as the number of cars in China increases.

How many people work at McDonalds in the world?

Strategy. Starting close to home, you could count the number of McDonalds in your town and ratio that number to the population of the rest of the United States. For example, Bloomington, Indiana, where I live, has four McDonald "restaurants" serving a population of about 120,000 people. Ratioing this to the U.S. population as a whole

$$\left(\frac{4\,\text{McD}}{1.2 \times 10^5 \text{ people}} \right) 2.95 \times 10^8 = 9800$$
restaurants in the United States

Based on local observations and questions of the people behind the counter, it seems that about 30 people work at each "restaurant"; hence,

$$\left(\frac{30\,\text{employees}}{\text{restaurant}}\right)\,9800\,\text{restaurants} \approx 3 \times 10^5\,\text{employees}$$

But this estimate is for the United States—what about the whole world? It is probably not right to use our factor of 3 (see just above) to link the United States' love for fast food to the rest of the world; for example, it is not likely that a Quarter-Pounder with Cheese will have the same appeal in India (1.3 billion people) as it does in the United States. Nevertheless, we could probably use a factor of 2 for this extrapolation and get about 600,000 McDonald employees worldwide. This estimate might be on the high side—Indiana has a relatively high concentration of McDonalds compared to other states. The truth seems to be that, in 2005, McDonalds had a total of 447,000 employees worldwide [*Fortune*, July 26, 2006, p. 122], so our estimate is not too bad.

How many American footballs can be made from one pig?

Strategy. Think about the size of a football—perhaps as a size-equivalent sphere—and about the size of a pig— perhaps as a big box—then divide one by the other. Let us assume that a football can be compressed into a sphere, and our best guess is that this sphere will have a diameter of about 25 cm (10 in.). Let us also imagine that a pig is a rectilinear box that is about 1 m long, 0.5 m high, and 0.5 m wide. This ignores the head, the tail, and the feet, which are probably not used to make footballs anyway.

$$\text{Pig area} = (4 \times 0.5 \times 1) + (2 \times 0.5 \times 0.5)\,\text{m}^2 = 2.5\,\text{m}^2$$

$$\text{Football area} = 4\pi r^2 = 4 \times 3.14 \times (25/2)^2$$
$$= 1963\,\text{cm}^2 \approx 2000\,\text{cm}^2$$

$$\text{Number of footballs} = \left(\frac{2.5\,\text{m}^2}{2000\,\text{cm}^2}\right)\left(\frac{10^4\,\text{cm}^2}{\text{m}^2}\right) \approx 10$$

This seems almost right, but most footballs are not made from pigskin any longer; like everything else, they are made from plastic.

1.3   IDEAL GAS LAW

We need this tool skill for dealing with many air pollution issues. The ideal gas law is

$$PV = nRT$$

where

$P =$ pressure in atmospheres (atm) or in Torr (remember 760 Torr = 1 atm)[2]
$V =$ volume in liters

---

[2] I know we should be dealing with pressure in units of Pascals (Pa), but I think it is convenient for environmental science purposes to retain the old unit of atmospheres—we instinctively know what that represents. For the purists among you, 1 atm $= 101,325$ Pa (or for government work, 1 atm $= 10^5$ Pa).

n = number of moles
R = gas constant (0.082 L atm/deg mol)
T = temperature in Kelvin (K = °C + 273.15)

The term mole refers to 6.02×10^{23} molecules or atoms; there are 6.02×10^{23} molecules or atoms in a mole. The term "moles" occurs frequently in molecular weights, which have units of grams per mole (or g/mol); for example, the molecular weight of N_2 is 28 g/mol. This number, 6.02×10^{23} (note the positive sign of the exponent), is known far and wide as Avogadro's number, and it was invented by Amadeo Avogadro in 1811.

We will frequently need the composition of the Earth's dry atmosphere; I have also included the molecular weight of each gas.

Gas	Symbol	Composition	Molecular weight
Nitrogen	N_2	78%	28
Oxygen	O_2	21%	32
Argon	Ar	1%	40
Carbon dioxide	CO_2	380 ppm	44
Neon	Ne	18 ppm	20
Helium	He	5.2 ppm	4
Methane	CH_4	1.5 ppm	16

The units "ppm" and "ppb" refer to parts per million and parts per billion, respectively. These are fractional units just like percent (%), which is parts per hundred. To get from the unitless fraction to these relative units, just multiply by 100 for %, by 10^6 for ppm, or by 10^9 for ppb. For example, a fraction of 0.0001 is 0.01% = 100 ppm = 100,000 ppb. For the gas phase, %, ppm, and ppb are all on a volume per volume basis (which is exactly the same as on a mole per mole basis); for example, the concentration of nitrogen in the Earth's atmosphere is 78 L of nitrogen per 100 L of air or 78 mol of nitrogen per 100 mol of air. It is *not* 78 g of nitrogen per 100 g of air. To remind us of this convention, sometimes these concentrations are given as "ppmv" or "ppbv." This convention applies to only gas concentrations—not to water, solids, or biota (where the convention is weight per weight).

What is the molecular weight of dry air?

Strategy. The value we are after should just be the weighted average of the components in air, mostly nitrogen at 28 g/mol and oxygen at 32 g/mol (and perhaps a tad of argon at 40 g/mol). Thus,

$$MW_{dry\ air} = 0.78 \times 28 + 0.21 \times 32 + 0.01 \times 40$$
$$= 29\ g/mol$$

What is the volume of 1 mole of gas at 1 atm and 0°C?

Strategy. We can just rearrange $PV = nRT$ and get

$$\left(\frac{V}{n}\right) = \left(\frac{RT}{P}\right) = \left(\frac{0.082\ L\ atm}{K\ mol}\right)\left(\frac{273\ K}{1\ atm}\right)$$
$$= 22.4\ L/mol$$

This value is 24.4 L/mol at 25°C. It will help us to remember both of these numbers, or at least, how to get from one to the other.

What is the density of the Earth's atmosphere at 0°C and 1 atm pressure?

Strategy. Remember that density is weight per unit volume, and we can get from volume to weight using the molecular weight, or in this case, the average molecular weight of dry air. Hence, rearranging $PV = nRT$

$$\frac{n(\text{MW})}{V} = \left(\frac{\text{mol}}{22.4\,\text{L}}\right)\left(\frac{29\,\text{g}}{\text{mol}}\right) = 1.3\,\text{g/L} = 1.3\,\text{kg/m}^3$$

What is the mass (weight) of the Earth's atmosphere?

Strategy. This is a bit harder, and we need an additional fact. We need to know the average atmospheric pressure in terms of weight per unit area. Once we have the pressure, we can multiply it by the surface area of the Earth to get the total weight of the atmosphere.

There are two ways to calculate the pressure: First, your average tire repair guy knows this to be 14.7 lb/in.^{2}, but we would rather use metric units.

$$P_{\text{Earth}} = \left(\frac{14.7\,\text{lb}}{\text{in.}^2}\right)\left(\frac{\text{in.}^2}{2.54^2\,\text{cm}^2}\right)\left(\frac{454\,\text{g}}{\text{lb}}\right)$$
$$= 1030\,\text{g/cm}^2$$

Second, remember from the TV weather reports that the atmospheric pressure averages 30 in. of mercury, which is 760 mm (76 cm) of mercury in a barometer. This length of mercury can be converted to a true pressure by multiplying it by the density of mercury, which is $13.5 \, \text{g/cm}^3$.

$$P_{\text{Earth}} = (76 \, \text{cm}) \left(\frac{13.5 \, \text{g}}{\text{cm}^3} \right) = 1030 \, \text{g/cm}^2$$

Next, we need to know the area of the Earth. I had to look it up – it is $5.11 \times 10^8 \, \text{km}^2$ – remember this! Hence, the total weight of the atmosphere is

$$\text{Mass} = P_{\text{Earth}} A = \left(\frac{1030 \, \text{g}}{\text{cm}^2} \right) \left(\frac{5.11 \times 10^8 \, \text{km}^2}{1} \right)$$

$$\times \left(\frac{10^{10} \, \text{cm}^2}{\text{km}^2} \right) \left(\frac{\text{kg}}{10^3 \, \text{g}} \right) = 5.3 \times 10^{18} \, \text{kg}$$

This is equal to 5.3×10^{15} metric tonnes.

Although this is not a realistic situation, it is useful to know what the volume (in L) of the Earth's atmosphere would be if it were all at 1 atm pressure and at 15°C (which is the average temperature of the lower atmosphere).

Strategy. Because we have just calculated the weight of the atmosphere, we can get the volume by dividing it by its density of $1.3 \, \text{kg/m}^3$, which we just calculated above.

$$V = \frac{\text{Mass}}{\rho} = 5.3 \times 10^{18} \, \text{kg} \left(\frac{\text{m}^3}{1.3 \, \text{kg}} \right) \left(\frac{288}{273} \right) \left(\frac{10^3 \, \text{L}}{\text{m}^3} \right)$$

$$= 4.3 \times 10^{21} \, \text{L}$$

Remember this number! Notice that the factor of 288/273 is needed to adjust the volume of air (1 m³ in the density) from 0°C (273 K) to 15°C (288 K)—the air gets warmer, so the volume increases.

An indoor air sample taken from a closed garage contains 0.9% of CO (probably a deadly amount). What is the concentration of CO in this air in units of g/m³ at 20°C and 1 atm pressure? CO has a molecular weight of 28.

Strategy. Given that the 0.9% amount is in moles of CO per moles of air, we need to convert the moles of CO to a weight, and the way to do this is using the molecular weight (28 g/mol). We also need to convert the moles of air to a volume, and the way to do this is using the 22.4 L/mol factor (corrected for temperature).

$$C = \left(\frac{0.9 \text{ mol CO}}{100 \text{ mol air}}\right)\left(\frac{28 \text{ g CO}}{\text{mol CO}}\right)\left(\frac{\text{mol air}}{22.4 \text{ L air}}\right)$$

$$\times \left(\frac{273}{293}\right)\left(\frac{10^3 \text{ L}}{\text{m}^3}\right) = 10.5 \text{ g/m}^3$$

Note the factor of 273/293 is needed to increase the volume of a mole of air from 0°C to 20°C.

1.4 STOICHIOMETRY

Chemical reactions always occur on an integer molar basis. For example,

$$C + O_2 \rightarrow CO_2$$

This means 1 mol of carbon (weighing 12 g) reacts with 1 mol of oxygen (32 g) to give 1 mol of carbon dioxide (44 g).

Here are a few atomic weights you should know.

H	1
C	12
N	14
O	16
S	32
Cl	35.5

Assume that gasoline can be represented by C_8H_{18}. How much oxygen is needed to completely burn this fuel? Give your answer in grams of oxygen per gram of fuel.

Strategy. First set up and balance the following combustion equation:

$$C_8H_{18} + 12.5\,O_2 \rightarrow 8\,CO_2 + 9\,H_2O$$

This stoichiometry indicates that 1 mol $(8 \times 12 + 18 = 114\,g)$ of fuel reacts with 12.5 mol $(12.5 \times 2 \times 16 = 400\,g)$ of oxygen to form 8 mol $(8 \times [12 + 2 \times 16] = 352\,g)$ of carbon dioxide and 9 mol $(9 \times [2 + 16] = 162\,g)$ of water. Hence, the requested answer is

$$\frac{M_{oxygen}}{M_{fuel}} = \left(\frac{400\,g}{114\,g}\right) = 3.51$$

This is called the stoichiometric ratio of the combustion system.

Assume that a very poorly adjusted lawn mower is operating such that the combustion reaction is $C_9H_{18} + 9\,O_2 \rightarrow 9\,CO + 9\,H_2O$. For each gram of fuel consumed, how many grams of CO are produced?

Strategy. Again we need to convert moles to weights using the molecular weights of the different compounds. The fuel has a molecular weight of 126 g/mol, and for every mole of fuel used, 9 mol of CO is produced. Hence,

$$\frac{M_{CO}}{M_{fuel}} = 9\left(\frac{28\,g}{mol}\right)\left(\frac{mol}{126\,g}\right) = 2.0$$

1.5 PROBLEM SET

1. Estimate the average spacing between carbon atoms in diamond, the density of which is 3.51 g/cm^3.
2. At Nikel, Russia, the annual average concentration of sulfur dioxide is observed to be 50 μg/m^3 at 15°C and 1 atm. What is this concentration of SO_2 in parts per billion?
3. Some modern cars do not come with an inflated spare tire. The tire is collapsed and needs to be inflated after it is installed on the car. To inflate the tire, the car comes with a pressurized can of carbon dioxide with enough gas to inflate three tires. What would this can of gas weigh? Assume an empty can weighs 0.2 kg.

4. The primary air quality standard for NO_2 in the United States, expressed as an annual average, is 53 ppb. What is the equivalent concentration in $\mu g/m^3$?

5. Atmospheric chemists love to use a gas concentration unit called "number density," in which the concentrations are given in units of molecules per cubic centimeter. Please calculate the number densities of oxygen in the atmosphere at sea level (1 atm, 25°C) and at an altitude of 30 km (0.015 atm, $-40°C$).

6. What would be the difference (if any) in the weights of two basketballs, one filled with air and one filled with helium? Please give your answer in grams. Assume the standard basketball has a diameter of 9.0 in. and is filled to a pressure of 8.0 lb/in.^{2} Sorry for the English units, but after all basketball was invented in the United States.

7. Acid rain was at one time an important point of contention between the United States and Canada. Much of this acid was the result of the emission of sulfur oxides by coal-fired electricity generating plants in southern Indiana and Ohio. These sulfur oxides, when dissolved in rainwater, formed sulfuric acid and hence "acid rain." How many metric tonnes of Indiana coal, which averages 3.5% sulfur by weight, would yield the H_2SO_4 required to produce a 0.9 in. rainfall of pH 3.90 precipitation over a 10^4 mile^2 area?

8. Although oil-fired electric power generating plants are becoming rare, let us assume one such plant consumes 3.5 million liters of oil per day, that the oil has an average composition of $C_{18}H_{34}$ and

density 0.85 g/cm^3, and that the gas emitted from the exhaust stack of this plant contains 45 ppm of NO. Please calculate the mass of NO emitted per day.

9. Imagine that 300 lb of dry sewage is dumped into a small lake, the volume of which is 300 million liters. How many tonnes of oxygen are needed to completely degrade this sewage? You may assume the sewage has an elemental composition of $C_6H_{12}O_6$.

10. Assume that an incorrectly adjusted lawn mower is operated in a closed two-car garage such that the combustion reaction in the engine is $C_8H_{14} + 15/2\ O_2 \rightarrow 8\ CO + 7\ H_2O$. How many grams of gasoline must be burned to raise the level of CO by 1000 ppm?

11. The average concentration of PCBs in the atmosphere around the Great Lakes is about 2 ng/m^3. What is this concentration in molecules/cm^3? The average molecular weight of PCBs is 320.

12. The following quote appeared in *Chemical and Engineering News* (September 3, 1990, p. 52): "One tree can assimilate about 6 kg of CO_2 per year or enough to offset the pollution produced by driving one car for 26,000 miles." Is this statement correct? Justify your answer quantitatively. Assume gasoline has the formula C_9H_{16} and that its combustion is complete.

13. If everyone in the world planted a tree tomorrow, how long would it take for these trees to make a 1 ppm difference in the CO_2 concentration? Assume that the world's population is 6 billion

and that 9 kg of O_2 is produced per tree each year regardless of its age. Remember CO_2 and H_2O combine through the process of photosynthesis to produce $C_6H_{12}O_6$ and O_2.

14. There are about 1.5×10^9 scrap tires in the world at the moment; this represents a major waste disposal problem. (a) If all of these tires were burned with complete efficiency, by how much (in tonnes) would the Earth's current atmospheric load of CO_2 increase? (b) Compare this amount to the current atmospheric CO_2 load. Assume rubber has a molecular formula of $C_{200}H_{400}$ and that each scrap tire weighs 8 kg, has a diameter of 48 cm, and is 85% rubber.

CHAPTER 2

MASS BALANCE

In environmental chemistry, we are often interested in features of a system related to time. For example, we might be interested in how fast the concentration of a pollutant in a lake is decreasing or increasing. Also, we might be interested in the delivery rate of some pollutant to an "environmental compartment" such as a lake or in how long would it take for a pollutant to clear out of a lake. These and other questions require us to master steady-state mass balance (in which the flow of something into an environmental compartment more or less equals its flow out of that compartment) and non-steady-state mass balance (in which the flow in does not equal the flow out). In these cases, an "environmental compartment" can be anything we want as long as we can define its borders and as long as we know something about what is flowing into and out of that system. For example, an environmental compartment can be the Earth's entire atmosphere, Lake Michigan, a bear, a house or garage, or the air "dome" over a large city.

These mass balance tools are usually quite familiar to engineers, but sometimes chemists and biologists do not pick up on this material in their undergraduate curricula. The ideas are simple but amazingly powerful.

Elements of Environmental Chemistry, by Ronald A. Hites
Copyright © 2007 John Wiley & Sons, Inc.

The following approach is derived from that presented by John Harte in his famous book *Consider a Spherical Cow*,[1] and the interested reader should see that little book for other examples of this approach.

2.1   STEADY-STATE MASS BALANCE

2.1.1 Flows, Stocks, and Residence Times

Let us imagine that we have some environmental compartment (e.g., a lake) with some water flow into it and some flow out of it (these could be two rivers in our lake example). Symbolically, we have

$$F_{in} \rightarrow \text{COMPARTMENT} \rightarrow F_{out}$$

where F = flow rate in units of amount per unit time (e.g., L/day for a lake). The total amount of material in the compartment is called the "stock" or sometimes the "burden." We will use the symbol M, which will have units of amount (e.g., the total liters of water in a lake). If $F_{in} = F_{out}$, then the compartment is said to be at "steady state."

The average time an item of the stock spends in the compartment is called the "residence time," and we will use the symbol τ (tau). It has units of time; for example, days. The reciprocal of τ is a rate constant with units of 1/time or time^{-1}; rate constants usually are represented by the symbol k.

---

[1] J. Harte, *Consider a Spherical Cow*, University Science Books, Sausalito, CA, 1988, 283 pp.

$$k = \frac{1}{\tau}$$

It is clear that

$$\tau = \left(\frac{M}{F_{in}}\right) = \left(\frac{M}{F_{out}}\right)$$

or

$$F = \left(\frac{M}{\tau}\right) = Mk$$

Remember this latter equation! Let us do some examples.

Imagine a one-car garage with a volume of 90 m³ and imagine that air in this garage has a residence time of 3.3 h. At what rate does the air leak into and out of this garage?

Strategy. The stock is the total volume of the garage, and the residence time is 3.3 h. Hence, the leak rate (in units of volume per unit time) is the flow through this compartment (the garage):

$$F = \left(\frac{M}{\tau}\right) = \left(\frac{90 \text{ m}^3}{3.3 \text{ h}}\right) = 27 \text{ m}^3/\text{h}$$

We can convert this to a flow rate in units of mass per unit time by using the density of air from the previous chapter, which is 1.3 kg/m^3.

$$F = \left(\frac{27 \text{ m}^3}{\text{h}}\right)\left(\frac{1.3 \text{ kg}}{\text{m}^3}\right) = 35 \text{ kg/h}$$

Methane (CH_4) is a greenhouse gas (more on this later), and it enters (and leaves) the Earth's atmosphere at a rate of about 500 million tonnes per year. If it has an atmospheric residence time of about 10 years, how much methane is in the atmosphere at any one time?

Strategy. Given the flow rate (F) and the residence time (τ), we can get the stock (M). Do not worry too much about remembering the appropriate equation; rather, worry about getting the units right.

$$M = F\,\tau = \left(\frac{500 \times 10^6\,\text{t}}{\text{year}}\right)\left(\frac{10\ \text{years}}{1}\right) = 5 \times 10^9\,\text{t}$$

Five billion tonnes of methane seems like an awful lot, but it is diluted by the entire atmosphere.

A big problem with this approach is that we often do not know the stock (M) in a compartment but rather we know the concentration (C) or that we know the stock but we really want to know the concentration. Let us define the concentration:

$$C = \left(\frac{M}{V}\right)$$

where V is the volume of the compartment. We can easily get the stock from the above equation by

$$M = C\,V$$

Hence,

$$F = \frac{CV}{\tau} = CVk$$

Given that the flow of oxygen into and out of the Earth's atmosphere is 3×10^{14} kg/year, what is the residence time of oxygen in the Earth's atmosphere?

Strategy. Because we are given the flow into and out of the compartment (the Earth's atmosphere), we need to know the stock (M) so that we can divide one by the other and get a residence time ($\tau = M/F$). Although we do not know the stock, we do know the concentration of oxygen in the atmosphere (21%). Thus, to calculate the stock of oxygen in the Earth's atmosphere, it is convenient to use the volume of the atmosphere at 15°C and at 1 atm pressure, which we figured out above to be 4.3×10^{21} L, and to multiply that by the concentration. Then, we just have to convert units to calculate the mass of oxygen in kg.

$$M = VC$$

$$= (4.3 \times 10^{21}\,\text{L})(0.21)\left(\frac{\text{mol}}{22.4\text{L}}\right)\left(\frac{273}{288}\right)\left(\frac{32\text{g}}{\text{mol}}\right)\left(\frac{\text{kg}}{10^3\text{g}}\right)$$

$$= 1.2 \times 10^{18}\,\text{kg}$$

Hence,

$$\tau = \left(\frac{M}{F}\right) = \left(\frac{1.2 \times 10^{18}\ \text{kg year}}{3 \times 10^{14}\,\text{kg}}\right)$$

$$= 0.40 \times 10^4\,\text{years} = 4000\,\text{years}$$

This suggests that the Earth's atmosphere is remarkably constant in terms of its oxygen concentration.

Carbonyl sulfide (COS) is present as a trace gas in the atmosphere at a concentration of 0.51 ppb; its

major source are the oceans, from which it enters the atmosphere at a rate of 6×10^8 kg/year. What is the residence time (in years) of COS in the atmosphere?

Strategy. Notice that the concentration is given as 0.51 ppb, which is a ratio of 0.51 L or mol of COS in 1 billion (10^9) L or mol of air. We can simplify this ratio to 0.51×10^{-9} and use this fraction in our calculation. If the units were parts per million, we could use it as a fraction equal to 0.51×10^{-6}, or if the units were percent, we could use it as a fraction equal to 0.0051.

Back to our problem. First, let us get the stock by multiplying the concentration by the volume of the atmosphere, and then we can divide that stock by the flow rate. We can do this all in one calculation:

$$\tau = \frac{CV}{F} = (0.51 \times 10^{-9})(4.3 \times 10^{21} \text{L}) \left(\frac{\text{year}}{6 \times 10^8 \text{kg}} \right)$$

$$\times \left(\frac{60 \text{ g COS}}{\text{mol}} \right) \left(\frac{\text{mol}}{22.4 \text{L}} \right) \left(\frac{273}{288} \right) \left(\frac{\text{kg}}{10^3 \text{ g}} \right) = 9.3 \text{ years}$$

Note that this is a much shorter residence time than that of oxygen.

Imagine that a particular Irish community has decided to dye the water of a small local lake (Lake Kelly) green and to keep it that way more or less permanently. (This is actually done with the Illinois River in Chicago, but only for St Patrick's Day.) The town fathers arrange for a green dye that is highly water soluble, nonvolatile, chemically stable, and nontoxic to be added to the lake at a

rate of 6.0 kg of the solid per day. The lake has a volume of 2.8×10^6 m^3, and the average water flow rate of the river feeding the lake is 6.9×10^3 m^3 day. Once the dye becomes well mixed in the lake, please estimate the dye's concentration in the lake's water.

Strategy. To make this problem tractable, let us forget about the possible evaporation of water from the lake's surface and assume that once the dye becomes well mixed in the lake, everything is at steady state. The latter statement means that the flow of the dye into the lake is balanced by its flow out of the lake in the water.

The water flow into and out of the lake is likely to be the same; this is almost always a good assumption unless there is major flooding of the surrounding countryside. Now we recall

$$C = \left(\frac{M}{V} \right)$$

$$M = F\tau$$

Hence,

$$C = \left(\frac{F\ \tau}{V} \right)$$

We know F and V; hence, we need the residence time of the pollutant in the lake. Because it is very water soluble, its residence time must be the same as the residence time of the water, which is given by

$$\tau = \left(\frac{M_{\text{water}}}{F_{\text{water}}} \right) = \left(\frac{2.8 \times 10^6\ \text{m}^3\ \text{day}}{6.9 \times 10^3\ \text{m}^3} \right) = 406\ \text{days}$$

Hence,

$$C = \left(\frac{F\tau}{V}\right) = \left(\frac{6.0\,\text{kg}}{\text{day}}\right)(406\,\text{days})$$

$$\times \left(\frac{1}{2.8 \times 10^6\,\text{m}^3}\right)\left(\frac{\text{m}^3}{10^3\,\text{kg}}\right)(10^6\,\text{ppm}) = 0.87\,\text{ppm}$$

Note that the density of water $(1000\,\text{kg/m}^3 = 1\,\text{g/cm}^3)$ is used here.

There is another way to do this problem. Note that

$$C = \left(\frac{V}{F_{\text{water}}}\right)\left(\frac{F_{\text{poll}}}{V}\right) = \left(\frac{F_{\text{poll}}}{F_{\text{water}}}\right)$$

which is just the dilution of the pollutant flow by the water flow. Hence,

$$C = \left(\frac{F_{\text{poll}}}{F_{\text{water}}}\right) = \left(\frac{6.0\,\text{kg/day}}{6.9 \times 10^3\,\text{m}^3/\text{day}}\right)$$

$$\times \left(\frac{\text{m}^3}{10^3\,\text{kg}}\right)(10^6\,\text{ppm}) = 0.87\,\text{ppm}$$

Note that this is 0.87 ppm on a weight per weight basis.

What if this same amount of dye was added in solution (rather than as a solid) and that the flow rate of the solution bringing this dye into the lake was $2.1 \times 10^3\,\text{m}^3/\text{day}$? In this case, what would the concentration be?

Strategy. In this case, the total flow rate of water would now be $6.9 \times 10^3\,\text{m}^3/\text{day}$ plus $2.1 \times 10^3\,\text{m}^3/\text{day}$ for a total of $9.0 \times 10^3\,\text{m}^3/\text{day}$. Hence, the residence time of water (and of the pollutant) would be

$$\tau = \left(\frac{M_{\text{water}}}{F_{\text{water}}}\right) = \left(\frac{2.8 \times 10^6 \text{ m}^3 \text{ day}}{9.0 \times 10^3 \text{ m}^3}\right) = 311 \text{ days}$$

And the concentration would be

$$C = \left(\frac{F\tau}{V}\right) = \left(\frac{6.0 \text{ kg}}{\text{day}}\right)(311 \text{ days})\left(\frac{1}{2.8 \times 10^6 \text{ m}^3}\right)$$

$$\times \left(\frac{\text{m}^3}{10^3 \text{ kg}}\right)(10^6 \text{ ppm}) = 0.67 \text{ ppm}$$

or by dilution

$$C = \left(\frac{F_{\text{poll}}}{F_{\text{water}}}\right) = \left(\frac{6.0 \text{ kg/day}}{(6.9 + 2.1) \times 10^3 \text{ m}^3/\text{day}}\right)$$

$$\times \left(\frac{\text{m}^3}{10^3 \text{ kg}}\right)(10^6 \text{ ppm}) = 0.67 \text{ ppm}$$

What if 10% of this water evaporated? Would this change the concentration of the dye in the lake? What would its concentration be?

Strategy. Because the dye does not evaporate with the water, the concentration would be higher in the water that is left behind. In effect, the flow rate for the water containing the dye is 90% of the total water flow rate. This changes the residence time for the water with the pollutant from 311 days to 346 days according to the following equation:

$$\tau = \left(\frac{M_{\text{water}}}{F_{\text{water}}}\right) = \left(\frac{2.8 \times 10^6 \text{ m}^3 \text{ day}}{9.0 \times 10^3 \text{ m}^3 \times 0.90}\right) = 346 \text{ days}$$

The concentration would then be

$$C = \left(\frac{F\tau}{V}\right) = \left(\frac{6.0\,\text{kg}}{\text{day}}\right)(346\,\text{days})\left(\frac{1}{2.8 \times 10^6\,\text{m}^3}\right)$$

$$\times \left(\frac{\text{m}^3}{10^3\,\text{kg}}\right)(10^6\,\text{ppm}) = 0.74\,\text{ppm}$$

We can also solve this problem by the dilution calculation we used above, in which the pollutant flow is diluted by the water flow. Hence,

$$C = \left(\frac{F_{\text{poll}}}{F_{\text{water}}}\right) = \left(\frac{6.0\,\text{kg/day}}{0.90 \times 9.0 \times 10^3\,\text{m}^3/\text{day}}\right)$$

$$\times \left(\frac{\text{m}^3}{10^3\,\text{kg}}\right)(10^6\,\text{ppm}) = 0.74\,\text{ppm}$$

which is the same as 0.67 ppm divided by 0.90. We know we have to divide rather than multiply because we know that the concentration has to go up if some of the water evaporated and the dye stays behind in the remaining water.

Note that these are all steady-state calculations. The concentration does not change as a function of time. Thus, this calculation cannot be used to determine the concentration of the dye in the lake just after the dye begins to be added or just after the dye stops being added. But do not worry; we have other tools for getting at these issues.

A sewage treatment plant is designed to process 9.3×10^6 L of sewage daily. What diameter (in feet) tank is required for the primary settling pro-

cess if the residence time is to be 7 h? Please assume the tank is cylindrical and 2 m deep.

Strategy. The only way to calculate the required size of the tank is to calculate the required area and then to calculate the diameter of the tank from that area. To calculate the area, we first need to calculate the necessary volume and then to divide that by the specified depth of 2 m. In this case, the volume is the amount of water (M in our notation) needed to give a residence time (τ) of 7 h with a flow rate (F) of 9.3×10^6 L/day.

$$V = F\tau = \left(\frac{9.3 \times 10^6 \, \text{L}}{\text{day}}\right)(7\,\text{h})\left(\frac{\text{day}}{24\,\text{h}}\right)\left(\frac{\text{m}^3}{10^3\,\text{L}}\right) = 2712\,\text{m}^3$$

$$A = V/h = \left(\frac{2712\,\text{m}^3}{2\,\text{m}}\right) = 1356\,\text{m}^2$$

Given that this is a circular tank, the radius of this tank would be

$$r = \sqrt{A/\pi} = \sqrt{\left(\frac{1356\,\text{m}^2}{3.14}\right)} = 20.78\,\text{m}$$

$$r = 20.78\,\text{m}\left(\frac{3.28\,\text{ft}}{\text{m}}\right) = 68\,\text{ft}$$

That is the radius, so the diameter is twice that of 136 ft, which is pretty big.

2.1.2 Adding Multiple Flows

If there are several processes by which something can be lost from a compartment, then each process has its own flow, which is given by the stock (M) times the rate constant (k) of that process. For example, let us imagine that we have one compartment with three processes by which some pollutant is leaving that compartment. The total flow out is

$$M_{total}k_{total} = M_1k_1 + M_2k_2 + M_3k_3$$

If each of the three processes applies to the same stock, namely to M_{total}, then $M_1 = M_2 = M_3 = M_{total}$, and the stocks cancel out giving us

$$k_{total} = k_1 + k_2 + k_3$$

Remembering that $k = 1/\tau$, we note that the total residence time is given by

$$\frac{1}{\tau_{total}} = \frac{1}{\tau_1} + \frac{1}{\tau_2} + \frac{1}{\tau_3}$$

We are back to that closed one-car garage (volume $= 90\,m^3$) with a badly adjusted lawn mower pumping out carbon monoxide at a rate of 11 g/h. Imagine that CO is lost from this garage by two processes: first, by simple mixing of clean air as it moves into and out of the garage, and second, by an unspecified chemical decay of the CO. Let us assume that the residence time of the air in the garage is 3.3 h and that the rate constant for the chemical decay of the CO is $5.6 \times 10^{-5}\,s^{-1}$. Under these conditions, what will be the average steady-state concentration of CO in this garage?

Strategy. There are two mechanisms by which CO is lost from this garage: First, by simple ventilation (or flushing) out of the garage with the normal air movement, and second, by chemical decay. The overall rate constant of the loss is the sum of the rate constants for these two processes. Remembering that the rate constant is the reciprocal of the residence time (and vice versa), we can find the ventilation rate constant by taking the reciprocal of the residence time. The chemical rate constant is given to be $5.6 \times 10^{-5}\,\text{s}^{-1}$. Thus, the overall rate constant is given by

$$k_{\text{overall}} = k_{\text{air}} + k_{\text{chem}} = \left(\frac{1}{3.3\,\text{h}}\right) + \left(\frac{5.6 \times 10^{-5}}{\text{s}}\right)$$

$$\times \left(\frac{60 \times 60\,\text{s}}{\text{h}}\right) = 0.303\,\text{h}^{-1} + 0.202\,\text{h}^{-1}$$

$$= 0.505\,\text{h}^{-1}$$

This calculation gives a residence time of

$$\tau_{\text{tot}} = \left(\frac{1\,\text{h}}{0.505}\right) = 1.98\,\text{h}$$

The concentration is given by the flow rate times the residence time of the CO divided by the volume of the garage:

$$C = \frac{M}{V} = \frac{F\tau}{V} = \left(\frac{11\,\text{g}}{\text{h}}\right)(1.98\,\text{h})\left(\frac{1}{90\,\text{m}^3}\right) = 0.242\,\text{g/m}^3$$

While we are at it, let us convert this to ppm units at 25°C.

$$C = \left(\frac{0.242\,\text{g}}{\text{m}^3}\right)\left(\frac{\text{mol}}{28\,\text{g}}\right)\left(\frac{24.4\,\text{L}}{\text{mol}}\right)\left(\frac{\text{m}^3}{10^3\,\text{L}}\right)(10^6\,\text{ppm})$$

$$= 210\,\text{ppm}$$

Occasionally, drinking water treatment plants will have "taste and odor" problems that result in a lot of complaints from their customers (after all, who wants smelly drinking water?). One compound that causes this problem is called geosmin.[2] Assuming this compound has a chemical degradation rate constant of $6.6 \times 10^{-3}\,\text{s}^{-1}$, at what flow rate could a treatment plant with a volume of $2500\,\text{m}^3$ be operated if a 10-min water contact time is required to remove qeosmin?

Strategy. The water contact time is the residence time of water in the plant, and from that time we can get a rate constant of $0.1\,\text{min}^{-1}$ for the flow of water. The chemical degradation rate constant is $6.6 \times 10^{-3}\,\text{s}^{-1}$, and thus, the total rate constant is

$$k_{\text{total}} = k_{\text{water}} + k_{\text{chemical}}$$

$$= \left(\frac{1}{10\,\text{min}}\right) + \left(\frac{6.6 \times 10^{-3}}{\text{s}}\right) \times \left(\frac{60\,\text{s}}{\text{min}}\right)$$

$$= 0.496\,\text{min}^{-1}$$

---

[2] For the organic chemistry aficionados among you, the structure of geosmin is

HO

Thus, the flow rate is given by

$$F = Mk = \left(\frac{2500\,\text{m}^3}{1}\right)\left(\frac{0.496}{\text{min}}\right)\left(\frac{60 \times 24\,\text{min}}{\text{day}}\right)$$

$$= 1.8 \times 10^6\,\text{m}^3/\text{day}$$

This is a typical flow rate of such a plant.

2.1.3 Fluxes are Not Flows!

The flow of something from a compartment *per unit area* of the surface through which the stuff is flowing is called a flux. We tend to use fluxes because they can easily be calculated from the concentration in a compartment times the speed at which that material leaves the compartment. This can be shown by

$$\text{Flux} = \left(\frac{\text{amount}}{\text{time} \times \text{area}}\right) = \left(\frac{\text{conc} \times \text{vol}}{\text{time} \times \text{area}}\right)$$

$$= \left(\frac{\text{conc} \times \text{area} \times \text{height}}{\text{time} \times \text{area}}\right)$$

$$= \left(\frac{\text{conc} \times \text{height}}{\text{time}}\right) = (\text{conc} \times \text{speed})$$

An example of the speed term in this equation is a deposition velocity, which is the rate at which a small particle leaves the atmosphere. Hence, the flux of, for example, lead from the atmosphere is

$$\text{Flux} = v_d\,C_p$$

where C_p is the concentration of lead on particles in atmosphere (e.g., in ng/m^3) and v_d is the velocity at which the particles fall out from the atmosphere (e.g., in cm/s). In terms of this example, the units of the flux are

$$\text{Flux} = \left(\frac{\text{cm}}{\text{s}}\right)\left(\frac{\text{ng}}{\text{m}^3}\right)\left(\frac{\text{m}}{10^2 \text{ cm}}\right) = \text{ng m}^{-2}\text{s}^{-1}$$

Another example where fluxes often show up is when dealing with so-called wet deposition from the atmosphere. In this case, "wet" deposition refers to both rain and snow, and the wet flux is the flow rate of some chemical coming out of the atmosphere with the rain and snow and depositing to soil or water per unit area. In this case, the velocity term is usually not given as the rate at which rain or snow falls through the atmosphere (this is much too variable), but rather the velocity term is integrated over a much longer time and is given as the rainfall rate in centimeters per year or occasionally per month. Thus,

$$\text{Flux} = \dot{p} \ C_{\text{wet}}$$

where C_{wet} is the concentration of, say, PCBs in rain or snow (e.g., in ng/L) and $\dot{p}$ is the rainfall rate (e.g., in cm/year; this notation is pronounced "p-dot"). In terms of this example, the units of the flux are

$$\text{Flux} = \left(\frac{\text{cm}}{\text{year}}\right)\left(\frac{\text{ng}}{\text{L}}\right)\left(\frac{\text{L}}{10^3 \text{ cm}^3}\right) = \text{ng cm}^{-2}\text{ year}^{-1}$$

By the way, one sees even experienced professionals get confused about the difference between a flow and a flux—these words cannot be used interchangeably. You should just remember that

$$\text{Flow} = \text{Flux} \times \text{Area}$$
$$\text{Flux} = \text{Flow/Area}$$

The North American Great Lakes are a major ecological, industrial, and recreational resource. It turns out that the average lead concentration in the air over Lake Erie is 11 ng/m^3, and the annual flow of lead from the atmosphere into this lake is 21,000 kg/year. Given that the area of Lake Erie is 25,700 km^2, please find the deposition velocity for lead into this lake.

Strategy. A little preliminary algebra goes a long way. We note that

$$\text{Flux} = \frac{F}{A} = C_p \, v_d$$

Rearranging the rightmost terms, we get

$$v_d = \frac{F}{A C_P} = \left(\frac{2.1 \times 10^4 \, \text{kg}}{\text{year}} \right) \left(\frac{1}{2.57 \times 10^4 \, \text{km}^2} \right)$$
$$\times \left(\frac{\text{m}^3}{11 \, \text{ng}} \right) \left(\frac{10^{12} \, \text{ng}}{\text{kg}} \right) \left(\frac{\text{km}^2}{10^6 \, \text{m}^2} \right) = 7.43 \times 10^4 \, \text{m/year}$$

$$v_d = \left(\frac{7.43 \times 10^4 \, \text{m}}{\text{year}} \right) \left(\frac{\text{year}}{60 \times 60 \times 24 \times 365 \, \text{s}} \right)$$
$$\times \left(\frac{10^2 \, \text{cm}}{\text{m}} \right) = 0.24 \, \text{cm/s}$$

This is a typical value.

Given the small mass of hydrogen molecules (H_2), it should not come as much of a surprise to note that the Earth has been losing H_2 at a flux of about 3×10^8 molecules cm^{-2} s^{-1} for most of its history. Please estimate the mass of hydrogen lost (in tonnes) from the atmosphere each year.

Strategy. Remembering that sometimes things are actually as simple as they look, this problem should be nothing more than multiplying the flux times the area and then converting units into mass.

$$F = \text{Flux} \times A = \left(\frac{3 \times 10^8 \, \text{molecules}}{\text{cm}^2 \, \text{s}} \right) \left(\frac{5.1 \times 10^8 \, \text{km}^2}{1} \right)$$
$$\times \left(\frac{10^{10} \, \text{cm}^2}{\text{km}^2} \right) \left(\frac{\text{mol}}{6.02 \times 10^{23} \, \text{molecules}} \right) \left(\frac{2 \, \text{g}}{\text{mol}} \right) \left(\frac{\text{t}}{10^6 \, \text{g}} \right)$$
$$\times \left(\frac{60 \times 60 \times 24 \times 365 \, \text{s}}{\text{year}} \right) = 1.6 \times 10^5 \, \text{t/year}$$

This seems like a lot, but apparently we have not run out yet.

Benzo[a]pyrene (BaP) is produced by the incomplete combustion of almost any fuel. This compound is one of the few proven human carcinogens. BaP enters Lake Superior by two mechanisms: wet deposition (rain and snow falling into the lake) and dry deposition (particles fall out to the surface of the lake). The concentration of BaP on particles in the air over Lake Superior is 5 pg/m³, and its concentration in rain over the lake is 2 ng/L. The area of Lake Superior is 8.21×10^4 km². What is the total flow of BaP from the atmosphere to Lake Superior?

Strategy for wet deposition. The total amount of BaP falling is the concentration in rain (C_r) times the amount of rain (V); the latter is the depth of rain times the area covered by the rain ($d \times A$). The flux is this amount of BaP divided by the area and by the time it took for that much rain to fall.

$$\text{Flux} = \frac{C_r A d}{A t} = C_r \dot{p}$$

Of course, this is the same equation we had before. The missing datum is the rainfall rate in depth per unit time or $\dot{p}$ in our notation. In the northeastern United States, this value is about 80 cm/year; in other words, if you stand out in a field for a year, and the rain did not drain away, the water level would be almost up to your waist. Once you have the flux, getting the flow is just the flux multiplied by the area.

$$F_{\text{wet}} = \text{Flux} \times A = C_r \dot{p} A$$

$$= \left(\frac{2\,\text{ng}}{\text{L}}\right) \left(\frac{80\,\text{cm}}{\text{year}}\right) \left(\frac{\text{m}}{10^2\,\text{cm}}\right)$$

$$\times \left(\frac{10^3\,\text{L}}{\text{m}^3}\right) \left(\frac{\text{kg}}{10^{12}\,\text{ng}}\right) \left(\frac{8.21 \times 10^4\,\text{km}^2}{1}\right)$$

$$\times \left(\frac{10^6\,\text{m}^2}{\text{km}^2}\right) = 130\,\text{kg/year}$$

Strategy for dry deposition. The flux is the deposition velocity times the atmospheric concentration of BaP on the particles. For a deposition velocity, let us use the number we calculated before, namely 0.24 cm/s.

Hence, the flow is

$$F_{\text{dry}} = \text{Flux} \times A = C_{\text{P}} v_{\text{d}} A$$

$$= \left(\frac{5\,\text{pg}}{\text{m}^3}\right) \left(\frac{0.24\,\text{cm}}{\text{s}}\right) \left(\frac{\text{m}}{10^2\,\text{cm}}\right) \left(\frac{\text{kg}}{10^{15}\,\text{pg}}\right)$$

$$\times \left(\frac{60 \times 60 \times 24 \times 365\,\text{s}}{\text{year}}\right) \left(\frac{8.21 \times 10^4\,\text{km}^2}{1}\right)$$

$$\times \left(\frac{10^6\,\text{m}^2}{\text{km}^2}\right) = 31\,\text{kg/year}$$

Hence, the total flux of BaP in this lake is the sum of the wet and dry fluxes

$$F_{\text{total}} = F_{\text{wet}} + F_{\text{dry}} = 130 + 31 = 160\,\text{kg/year}$$

2.2 NON-STEADY-STATE MASS BALANCE

What happens if the flow into a compartment is not equal to the flow out of a compartment? For example, when a pollutant just begins flowing into a clean lake, the lake is not yet at steady state. It takes some time for the flow out of the lake to increase to the new, higher (but constant) concentration of the pollutant. The time it takes to come up to this new concentration depends on the residence time of the pollutant in the lake. Long residence times imply that it takes a long time for the pollutant to come to the new concentration.

The differential equation describing all this is simply

$$\frac{\mathrm{d}M}{\mathrm{d}t} = F_{in} - F_{out}$$

where M is the mass of pollutant (or anything else) in the lake and t is time.[3] When $F_{in} = F_{out}$, $\mathrm{d}M/\mathrm{d}t$ is zero, and time is not a factor. Hence, the system is at steady state, and we just covered this condition. When the flows are not equal, then time *is* a factor, and we need to solve this differential equation.

2.2.1 Up-Going Curve

First, let us consider a situation in which the concentration starts at zero and goes up to a steady-state value. An example is a clean lake in which someone starts dumping some pollutant at a constant rate (perhaps the green dye mentioned above). Because the dumping just started, the flow of the pollutant into the lake is greater than its flow out of the lake, and the flow into the lake is constant. We know that the concentration at time $= 0$ is

---

[3] Please do not be frightened by calculus. In this equation, d refers to difference, and it can be read as the difference in the amount of a pollutant divided by the difference per unit time. In other words, this is the flow rate at which the amount of the pollutant increases or decreases. A common derivative that we are all used to dealing with is $\mathrm{d}x/\mathrm{d}t$, which is the change in distance per unit time. This is called speed, and every car in the world is equipped with a gauge to measure this derivative. In fact, many political units make money by collecting fines if your car exceeds specified values of this derivative and if the policeman catches you. The posted limit of this derivative on the Eisenhower Interstate System is usually 65 or 70 miles/h.

zero and at some very long time, the concentration is equal to the steady-state concentration, but we need to know the shape of the curve in between. We can get at this using a simple differential equation.

Let us assume that the pollutant's flow rate out of the lake is proportional to the amount of pollutant in the lake. That is, when the pollutant's concentration in the lake is low, the pollutant's flow out of the lake is low, and when the pollutant's concentration is high, the pollutant's flow out of the lake is high. This is usually a good assumption. The proportionality constant is a rate constant (k) with units of reciprocal time. Hence,

$$F_{out} = kM = M/\tau$$

Thus, the differential equation becomes

$$\frac{dM}{dt} = F_{in} - kM$$

where F_{in} is fixed or constant. Dividing both sides of this equation by V, the volume of the lake, we get

$$\frac{dC}{dt} = \frac{F_{in}}{V} - kC$$

where C is concentration (amount per unit volume). We note that at time $= 0$, the concentration $= 0$; and at time $= \infty$, the concentration $= C_{max}$, which is the steady-state concentration. We can now solve this differential equation and get

$$C = C_{max}[1 - \exp(-kt)]$$

Notice that $\exp(x)$ is the same as e^x, where e is 2.718,282. Find the button on your calculator now.

Also, remember that $\ln(e^x) = x$. We will call this the "up-going" equation, and you should remember it. We are going to use logarithms a lot; for those of you who have forgotten about them, please learn.[4]

A plot of this equation is shown in Figure 2.1. Note that concentration goes up with time and levels off at a maximum value and that this maximum value is the steady-state concentration called C_{max}, which is assumed to be 2 ppm in this example.

Let us check some signposts along this curve.

How long will it take for the concentration to get one half of the way to its steady state concentration?

Strategy. Call this time $t_{1/2}$. At this time, $C = C_{max}/2$. In the up-going equation, we have

$$C_{max}/2 = C_{max}[1 - \exp(-k\,t_{1/2})]$$

---

[4] A logarithm is the power to which some "base number" needs to be raised to get the number you are after in the first place. The "base number" is usually 10 or e (2.7182...). Let us first focus on 10 as a base. Let us assume that the number you have is 100. We know that $10^2 = 100$; thus, the logarithm of 100 is 2. We usually write this as $\log(100) = 2$. In the same way, the logarithm of 1 million is 6. Clearly, the logarithm of 2 is between 0 ($10^0 = 1$) and 1 ($10^1 = 10$), and it turns out that the logarithm of 2 is 0.301. Logarithms can also be negative. For example, $\log(0.01) = -2$ and $\log(0.005) = -2.30$. If the base is e, the abbreviation is ln (called the natural logarithm) not log (called the common logarithm). The numbers are, of course, different, but the same idea applies. For example, it turns out that $e^{2.303} = 10$; thus, the natural logarithm of 10 is 2.303 or $\ln(10) = 2.303$. The logarithm of zero is impossible because you cannot raise any number to any power and get 0. Remember the inverse function of a logarithm is exponentiation. Your calculator will supply logarithms and exponents with either base number.

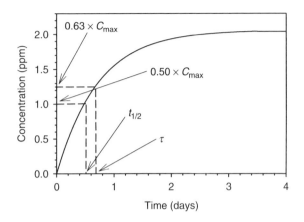

Figure 2.1   Concentration of some pollutant as a function of time in an environmental compartment that had none to start and to which the pollutant was added at a constant rate such that a steady state was achieved after a sufficient time. In this example, the steady-state concentration is 2 ppm and the rate constant is $1.4\,\mathrm{day}^{-1}$. Two signposts on this curve are shown.

Canceling $C_{\max}$ from both sides and solving for $t_{1/2}$, we get

$$1/2 = \exp(-k\,t_{1/2})$$

Taking the natural logarithm of both sides, we get

$$\ln(1) - \ln(2) = -k\,t_{1/2}$$

Because $\ln(1) = 0$

$$t_{1/2} = \ln(2)/k$$

where $\ln(2)$ is 0.693 (see your calculator—note that $e^{0.693} = 2$, which is the very definition of a logarithm). I beg of you, please remember this equation! Note that

$k = 1/\tau$; hence,

$$t_{1/2} = \tau \ln(2)$$

In the example shown in the plot, $k = 1.4\,\text{day}^{-1}$; hence, $t_{1/2} = 0.5$ days (see the above plot).

To what fraction of the steady-state concentration will the environmental compartment get in the residence time of the compartment?

Strategy. At $t = \tau$, a slight rearrangement of the up-going equation gives

$$C/C_{\text{max}} = 1 - \exp(-k\,\tau)$$

But of course, $k\,\tau = 1$; hence,

$$C/C_{\text{max}} = 1 - \exp(-1) = 1 - 0.368 = 0.632$$

which is about two thirds of the way to C_{max}. See the above plot.

Back to the bad lawn mower, which is being operated in a closed shed with a volume of 8 m³. The engine is producing 0.7 g of CO per minute. The ventilation rate of this shed is 0.2 air changes per hour. Assume that the air in the shed is well mixed and that the shed initially had no CO in it. How long would it take for the CO concentration to get to 8000 ppm?

Strategy. Obviously, we will use the up-going equation. To use this equation, we need to know the steady-state concentration (C_{max}). Once we have this number, we can use the given values of C (8000 ppm) and

$k\,(0.2\,\mathrm{h}^{-1})$ in the equation to get the time (t). The steady-state or maximum concentration is given by

$$C_{\max} = \frac{F\tau}{V} = \frac{F}{Vk}$$

$$= \left(\frac{0.7\,\mathrm{g}}{\min}\right)\left(\frac{1}{8\,\mathrm{m}^3}\right)\left(\frac{\mathrm{h}}{0.2}\right)\left(\frac{60\,\min}{\mathrm{h}}\right)\left(\frac{\mathrm{mol}}{28\,\mathrm{g}}\right)$$

$$\times \left(\frac{24.4\,\mathrm{L}}{\mathrm{mol}}\right)\left(\frac{\mathrm{m}^3}{10^3\,\mathrm{L}}\right)(10^6\,\mathrm{ppm}) = 22{,}900\,\mathrm{ppm}$$

$$C = C_{\max}[1 - \exp(-kt)]$$

$$8000 = 22900\,[1 - \exp(-0.2\,t)]$$

$$\frac{8000}{22{,}900} - 1 = -0.65 = -\exp(-0.2\,t)$$

$$\ln(0.65) = -0.2\,t$$

$$\frac{0.430}{0.2} = t = 2.15\,\mathrm{h}$$

These calculations are strung out in detail to demonstrate the use of logarithms and exponentiation. For simplicity, we have been a bit naughty by not writing all of the units as we proceeded, but it is clear that the concentrations (C and $C_{\max}$) have to be in the same units so that they will cancel and that the units of k and t have to be the reciprocal of one another so that they will vanish. In this case, the answer is in hours. So this calculation says that it will take a bit over 2 h for the concentration in the shed to get to 8000 ppm, but we do not suggest that one should wait in the shed to see if this is correct.

2.2.2 Down-Going Curve

What if the input rate is zero ($F_{in} = 0$) and the output rate is proportional to the amount in the environmental compartment, that is $F_{out} = kC$? In this case, the concentration in the compartment would go down with time and the differential equation would be

$$\frac{dM}{dt} = F_{in} - F_{out} = -F_{out} = -kM$$

or we can put this in concentration units

$$\frac{dC}{dt} = -\frac{F_{out}}{V} = -\frac{M}{\tau V} = -kC$$

This can be rearranged to give

$$\frac{dC}{C} = -k\, dt$$

Assuming that at time $= 0$, the concentration is C_0; the solution to this differential equation is

$$\ln(C) = -kt + \ln(C_0)$$

which is a straight line with a slope of $-k$. We can also exponentiate both sides of this equation and get

$$C = C_0 \exp(-kt)$$

We will call this the "down-going" equation. Remember this! A plot of this equation is shown in Figure 2.2.

The most important signpost along this curve is the half-life, which is the time it takes for half of the

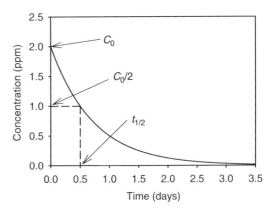

Figure 2.2 Concentration of some pollutant as a function of time in an environmental compartment that had a concentration of C_0 to start and to which no more pollutant was added. In this example, the initial concentration (C_0) is 2 ppm and the rate constant is $1.4\,\text{day}^{-1}$. The half-life of 0.5 days is shown on this curve.

compound or material to disappear from the compartment. In this case,

$$1/2\,C_0 = C_0\exp(-k\,t_{1/2})]$$

Hence,

$$t_{1/2} = \ln(2)/k$$

This is just like that for the up-going curve. Note the units of k and $t_{1/2}$. Remember this equation too! Actually do not remember the equation, just remember the units of k and τ and $\ln(2)$. In this example,

$$t_{1/2} = \frac{\ln(2)}{1.4\,\text{day}^{-1}} = 0.5\,\text{day}$$

In the course of producing nuclear weapons, an unnamed country (often in trouble with the United Nations) had a small spill of promethium-147 (^{147}Pm) in September 1989. This spill totaled 4.5 microCuries (µCi), covered a soil area of $5\,m^2$, and penetrated to a depth of 0.5 m. In August 1997, the United Nations tested this site, and the concentration of ^{147}Pm was found to be $0.222\,\mu Ci/m^3$ of soil. What is the half-life (in years) of ^{147}Pm?

Strategy. This is obviously a time to use the down-going equation, and thus, we need to know C_0 in September 1989 and C in August 1997. This is an elapsed time of 95 months, so t in our equation will be 95 months. The concentration units can be anything we want because they will cancel out during the calculation.[5] So to put the concentrations in consistent units, we need to find the volume of the affected area and divide it into the initial radioactivity.

$$C_0 = \left(\frac{\text{Radioactivity}_0}{\text{Area} \times \text{height}}\right) = \left(\frac{4.5\,\mu\,Ci}{5\,m^2 \times 0.5\,m}\right)$$

$$= 1.8\,\mu\,Ci/m^3$$

Given that we know C in the same units 95 months later, we can now solve for k.

---

[5] By the way, ds in differential equations do not cancel because they are not variables, but rather they are operators, meaning in this case that we are to take the difference before we take the ratio.

$$C = C_0 \exp(-kt)$$
$$0.222 = 1.8 \ \exp(-95k)$$
$$\ln\left(\frac{0.222}{1.8}\right) = -2.093 = -95k$$
$$k = 0.0220 \, \text{month}^{-1} \left(\frac{12 \, \text{months}}{\text{year}}\right) = 0.264 \, \text{year}^{-1}$$
$$t_{1/2} = \frac{\ln(2)}{k} = \frac{\ln(2)}{0.264} = 2.62 \, \text{years}$$

This is exactly what it should be, but of course, it should agree perfectly, given that I made up the data.

2.2.3 Working with Real Data

You will very often need to find a rate constant or a half-life from real data. In the old days (before Microsoft), it was difficult to plot these data as a curved line as shown in Figure 2.2 and to fit a curved line to it such that one could read off the rate constant, k. Thus, it was handy to convert these data into a linear form and to fit a straight line. We can do this easily by taking the natural logarithm of the down-going equation and getting

$$\ln \ C = \ln \ C_0 - k \, t$$

Hence, a plot of $\ln(C)$ versus t will give a straight line with a slope of $-k$ and an intercept of $\ln(C_0)$. Because each number will have some measurement error, you will need to use statistical regression techniques to get the best values of C_0 and k. The technique is simple: Just convert the concentrations to their natural logarithms and fit a straight line to a plot of the values versus time. Often, you can eyeball a line or use the linear regression feature of your calculator.

Now that everyone has a powerful computer in their back-pack or on their desk, it is simple to fit a straight or even a curved line to data using the TrendLine feature of Excel. As an example, let us work on some real data for DDT in trout from Lake Michigan.

Year	Concentration (ppm)
1970	19.19
1971	13.00
1972	11.31
1973	9.96
1974	8.42
1975	7.50
1976	5.65
1977	6.34
1978	4.58
1979	6.91
1980	4.74
1981	3.22
1982	2.74
1984	2.22
1986	1.10
1988	1.44
1990	1.39
1992	1.16
1995	0.98
1998	0.85

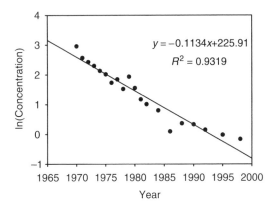

Figure 2.3 Graph created with Microsoft Excel showing the natural logarithms of the concentrations of DDT in trout from Lake Michigan (see the above table) as a function of time and showing a fitted straight line (using the TrendLine feature). The negative slope of this line is the rate constant.

It is straightforward to enter these values in an Excel spreadsheet with the years in column A and the concentrations in column B—please do this now. Next, you can take the natural logarithms of the concentrations in column C using the built-in LN function. Now, if you plot column A (as the x-values) versus column C (as the y-values) using the Chart Wizard and using the XY (scatter) chart type, you should get a plot that looks like the graph in Figure 2.3. Now, right-click on the data and select TrendLine. Select the linear TrendLine—be sure to go to the options tab and ask to have the equation and the correlation coefficient displayed on the chart. Now, you will have a plot of $\ln(C)$ versus time with a fitted straight line as shown in Figure 2.3. The statistical

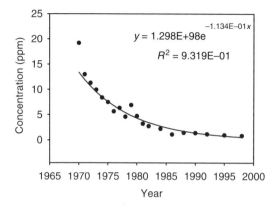

Figure 2.4 Graph created with Microsoft Excel showing the concentrations of DDT in trout from Lake Michigan (see the above table) as a function of time and showing a fitted exponential line (using the TrendLine feature). The negative exponent is the rate constant.

results are $r^2 = 0.932$ and slope $= -0.1134 \, \text{year}^{-1}$. The slope is $-k$ and $t_{1/2} = \ln(2)/0.1134 = 6.1$ years.

It turns out that we do not have to go through the effort of taking the natural logarithms ourselves. If we just plot the concentrations (not their logarithms) versus time, we can select the exponential TrendLine from the menu and get an exponential curve fitted to our data; see Figure 2.4. Note that the r^2 values and the k values are exactly the same regardless of whether we use the linear fit or the exponential fit. This is because Excel converts the concentrations internally to their logarithms, fits a straight line to these converted data, and then converts everything back to nonlogarithmic units. Thus, you might as well use the exponential curve fit feature if you have Excel available. If all you have is a paper and pencil (and calculator), then you may want to convert the concentrations to their natural logarithms, plot them versus time, and eyeball a straight line to these converted data. In any case, always plot your data.

The concentrations of octachlorostyrene in trout in the Great Lakes have been measured over the years with the following results: 1986, 26 ppm; 1988, 18; 1992, 13; 1995, 12; 1998, 6.2; and 2005, 1.8. What is your best estimate of this compound's half-life (in years) in these fish?

Strategy. Let us just plot the raw data in Excel and use the TrendLine feature to fit an exponential line. Be sure to turn on the optional "show equation and correlation coefficients" feature. We get the plot shown in Figure 2.5.

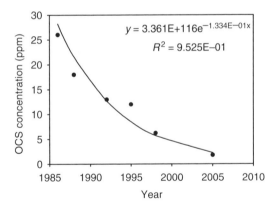

Figure 2.5   Graph created with Microsoft Excel showing the concentrations of octachlorostyrene in trout from the Great Lakes (see the above paragraph) as a function of time and showing a fitted exponential line (using the Trend-Line feature). The negative exponent is the rate constant.

Notice that the rate constant is $0.1334\,year^{-1}$. Thus, the half-life is

$$t_{1/2} = \frac{\ln(2)}{0.1334} = 5.2\,years$$

For the up-going equation, one cannot do a simple linearization like we could for the down-going equation. In this case, the equation can be made linear if, and only if, C_{max} is known or can be assumed. In this case,

$$k = \frac{\ln\left(\frac{C_{max}}{C_{max}-C}\right)}{t}$$

Please verify for yourself that this equation is correct. If one does not know C_{max} or cannot make a reasonable guess, then one must use nonlinear curve fitting techniques, the simplest of which is the Solver tool in Excel.

Let us go back to the problem of the green dye dumped into Lake Kelly. Before the town fathers decided to dye the lake water green, the concentration of dye was zero. After they started dumping the dye into the lake, the concentration started to increase; in fact, the concentrations (in ppm) measured every 6 months were 0.33, 0.50, 0.66, and 0.70.

Strategy. We know from the previous problem that the steady-state concentration of the dye was $0.87\,ppm$. This is the same as C_{max} in the above equation. Thus, we can calculate k for each time measurement from the above equation.

Time (years)	C (ppm)	Calculated k (year^{-1})
0.5	0.33	0.954
1.0	0.50	0.855
1.5	0.66	0.948
2.0	0.70	0.816

Notice that the calculated k values differ from one another and range from 0.816 to 0.954; this is because of the concentration measurement error associated with each value. To reduce these results to one manageable value, it is reasonable to take the average and standard deviation of these four values.[6] The resulting values are (0.893 ± 0.069) year^{-1}, which converts to a resident time of 410 days, which is virtually the same as the value we calculated previously for the green dye in this lake.

2.2.4 Second-Order Reactions

The rates of most of the processes we have been discussing depend on the amount of only one thing in a compartment. Two examples are as follows: (a) The rate at which radon is removed from a house depends only on the concentration of radon in that house. (b) The rate at which a pollutant is removed from a lake depends only on the amount of that pollutant in the lake $(-\mathrm{d}C/\mathrm{d}t = k\,C)$. These are all called *first-order* processes because they depend on one thing.

---

[6] See your calculator on how to do this.

Some environmental reactions require two reactants, and the rate of that reaction depends on the concentrations of *both* of the reactants. These are known as *second-order* reactions. For example, the reaction of the hydroxyl radical (OH) with naphthalene (Nap)[7] in the atmosphere

$$Nap + OH \rightarrow products$$

takes place in the gas phase, and the rate of the reaction depends on the concentrations of both naphthalene and OH. Another example might be some pollutant that reacts slowly with water once it gets into a river or lake

$$Crud + H_2O \rightarrow products$$

In these cases, the rate at with the pollutant (naphthalene or crud) disappears from the compartment is a function of *both* the concentrations of the pollutant and of the other chemical species. Thus, in our second example, the rate law is

$$-\frac{d[crud]}{dt} = k_2[crud][H_2O]$$

where k_2 is a second-order rate constant. Solving this differential equation is slightly tricky if the concentrations of both chemical species are changing with time. Fortunately, one reactant is often present in great excess

---

[7] The structure of naphthalene (note the two *h*s in the spelling) is

(in our examples, OH in the atmosphere or water in a river), and as a result, we can assume that its concentration does not change over time. Thus, we can combine the constant concentration with the second-order rate constant to get a pseudo-first-order rate constant. In our first example, this would be

$$-\frac{d[\text{Nap}]}{dt} = k_2[\text{OH}][\text{Nap}] = k'[\text{Nap}]$$

where k_2' is the pseudo-first-order rate constant. The advantage of this approach is that we can use everything we learned from first-order kinetics with a second-order system.

Let us look at this naphthalene reaction in detail. The second-order rate constant for this reaction is

$$k_2 = 24 \times 10^{-12}\,\text{cm}^3/\text{s}$$

Note that the units of this rate constant are reciprocal concentration (in number density units) and reciprocal time (in s^{-1}). Remember that "number density" is a fancy way of saying molecules per unit volume (usually as molecules/cm^3).

It turns out that the concentration of OH is almost always about $9.5 \times 10^5\,\text{cm}^{-3}$. Thus, for the reaction of naphthalene with OH, the pseudo-first-order rate constant is

$$k' = k_2[\text{OH}] = \left(\frac{24 \times 10^{-12}\,\text{cm}^3}{\text{s}}\right)\left(\frac{9.5 \times 10^5}{\text{cm}^3}\right)$$
$$= 2.28 \times 10^{-5}\,\text{s}^{-1}$$

Note that this now has the units of a first-order rate constant (namely reciprocal time).

We can use this pseudo-first-order rate constant to make some useful estimates. For example, the residence time of naphthalene in the atmosphere due to OH reactions is

$$\tau = \frac{1}{k'} = \left(\frac{s}{2.28 \times 10^{-5}}\right)\left(\frac{h}{60 \times 60\,s}\right) = 12\,h$$

In general, converting second-order rate constants, which are hard to deal with, into pseudo-first-order rate constants, which are easier to deal with, is a common strategy in environmental chemistry.

This is important, so we will repeat it: You can almost always convert a second-order rate constant into a first-order rate constant by multiplying the second-order constant by the concentration of one of the reactants if the concentration of that reactant does not change much during the reaction.

The average global concentration of atmospheric methane is 1.74 ppm (at 0°C and 1 atm), and the second-order rate constant for the reaction of CH_4 and OH is 3.6×10^{-15} cm^3/s. What is the rate (in Tg/year) of methane destruction by this reaction? Assume [OH] is always 10^6 cm^{-3}.

Strategy. First let us calculate the pseudo-first-order rate constant and then use it to get the flow rate using the "normal equations."

$$k' = k_2[\text{OH}] = \left(\frac{3.6 \times 10^{-15}\,\text{cm}^3}{\text{molecules s}}\right)\left(\frac{10^6\,\text{molecules}}{\text{cm}^3}\right)$$
$$= 3.6 \times 10^{-9}\,\text{s}^{-1}$$
$$F = \frac{M}{\tau} = kCV = \left(\frac{3.6 \times 10^{-9}}{\text{s}}\right)\left(\frac{1.74\,\text{L CH}_4}{10^6\,\text{L air}}\right)$$
$$\times \left(\frac{4.3 \times 10^{21}\,\text{L}}{1}\right)\left(\frac{273}{288}\right)\left(\frac{16\,\text{g}}{\text{mol}}\right)\left(\frac{60 \times 60 \times 24 \times 365\,\text{s}}{\text{year}}\right)$$
$$\times \left(\frac{\text{Tg}}{10^{12}\,\text{g}}\right) = 12900\,\text{Tg/year}$$

This may seem like a lot, but remember there are a lot of sources of methane to the atmosphere, including methanogenic bacteria.

2.3   PROBLEM SET

1. *USA Today* (September 30, 1996) ran a story about a large tire dump in Smithfield, Rhode Island. Fourteen acres of the site are covered by tires to an average height of 25 ft. (a) How many tires are in this dump? (b) Assuming that a tire lasts about 3 years on the typical car, estimate the fraction of the United States' annual tire production that is in this dump. Hint: 1 acre = 4047 m^2.

2. This is a true story. A lawyer in Bloomington wanted to impress a judge with the high number of PCB molecules that a typical person takes in with each breath. He called Professor Hites for this fact, but he did not have this information at hand. However, one of his students had recently determined

that the flux of PCBs from the atmosphere to Bloomington's surface averaged $50 \, \mu g \, m^{-2} \, year^{-1}$. You may assume that the area of Bloomington is 100 square miles, that the deposition velocity of PCBs is 0.3 cm/s, and that the molecular weight of PCBs is 320. What would you tell the lawyer?

3. Assume a lake is fully stratified and that a pollutant enters the upper layer from a river at a rate of 35 kg/year, and it enters the lower layer from ground water seepage at 4 kg/year. Because of sedimentation, the residence time of the pollutant in the lower layer is 1.5 years. The average concentration in the whole lake is 80 ng/L, the total lake volume is $10^9 \, m^3$, and everything is at steady state. (a) Draw a diagram of the system. (b) What is the total amount of pollutant in the lake? (c) Set up equations relating the stocks, flows, and residence times (define your terms on the diagram). (d) Solve for the residence time in the upper layer.

4. PCBs are transported through the environment in the vapor phase to Lake Superior. Let us assume that the flux to this lake is almost all due to rainfall, which has an average PCB concentration of 30 ng/L. The average depth of Lake Superior is 150 m, and the rainfall averages 80 cm/year. Assume that everything has been at steady state over the last several decades and that the residence time of PCBs in the lake is 3 years. For the moment, do not worry about deposition of PCBs to the sediment of this lake. What is the concentration of PCBs in Lake Superior water?

5. The PCB concentration in the air inside a typical home is 400 ng/m^3, and the ventilation rate is such that it takes, on average, 10 h to change the indoor air. If the indoor PCBs are coming from a leaking capacitor, how much leaks per year?

6. Although this number is probably now decreasing, a few years ago, nearly 40% of the 110 million new bicycles manufactured annually in the world were produced and used in China. How often (in years) did the average citizen of China get a new bicycle?

7. A grizzly bear eats 20 fish per day, and a pollutant at a concentration of 30 ppb contaminates the fish. On average, the pollutant remains in the bear's body for 2 years. What is the steady-state concentration (in ppm) of the pollutant in the bear?

8. A one-compartment home of volume 400 m^3 has an infiltration rate of 0.3 air changes per hour with its doors and windows closed. During an episode of photochemical smog, the outdoor concentration of PAN is 75 ppb. If the family remains indoors, and the initial concentration of PAN inside is 9 ppb, how long will it be before the PAN concentration inside rises to 40 ppb?

9. A soluble pollutant is dumped into a clean lake starting on day 0. The rate constant of the increase is $0.069\,\text{day}^{-1}$. (a) Sketch a plot of the relative concentration from day 0 to day 60; be sure to label the axes with units and numbers. (b) What fraction of the steady-state concentration is reached after 35 days?

10. You have been asked to measure the volume of a small lake. You dump in 5.0 L of a 2.0 M solution of a dye, which degrades with a half-life of 3.0 days. You wait exactly 1 week for the lake to become well mixed (during this time, assume no water is lost); you then take a 100 mL sample. The dye's concentration in this sample is 2.9×10^{-6} M. What is the lake's volume? Remember "M" is moles per liter.

11. A pollutant is dumped into a clean lake at a constant rate starting on July 1, 1980. When the pollutant's concentration reaches 90% of its steady-state value, the flow of the pollutant is stopped. On what date will the concentration of the pollutant fall to 1% of its maximum concentration? Assume that the rate constants of the increase and decrease are both 0.35 year^{-1}.

12. The residence time of the water in Lake Erie is 2.7 years. If the input of phosphorus to the lake is halved, how long will it take for the concentration of phosphorus in the lake water to fall by 10%?

13. DDT was applied to a field that was then plowed twice. The initial concentration was 49 ppm. The concentration of DDT in the soil was then measured at monthly intervals; the results were 36, 26, 18, 14, 10, 7.3, 5.5, and 3.9 ppm. What is the half-life of DDT in this particular case?

14. The atmospheric reaction $NO_2 + O_2 \rightarrow NO + O_3$ is first order with respect to NO_2; it has a rate constant of 2.2×10^{-5} s^{-1} at 25°C. Assuming no new inputs of NO_2, what percent of NO_2 would be remaining after 90 min?

15. Assume that the following data are the average concentrations (in ppq, 10^{-15} parts) of 2378-TCDD equivalents in people from the United States, Canada, Germany, and France in the given years: 1972, 19.8; 1982, 7.1; 1987, 4.1; 1992, 3.2; 1996, 2.1; and 1999, 2.4 ppt. [L.L. Aylward and S.M. Hayes, *Journal of Exposure Analysis and Environmental Epidemiology*, **12**, 319–328 (2002)]. What is the half-life of 2378-TCDD in these people? What will be the concentration in 2008?

16. Kelthane degrades to dichlorobenzophenone, DCBP. Kelthane is difficult to measure, but DCBP is easier. In an agricultural experiment similar to that described above, the following data (time in weeks, concentration in ppm) were obtained: 6, 19; 9, 26; 12, 32; 14, 36; 18, 41; 22, 45; 26, 48; 30, 50; and 150, 58. What is the half-life of Kelthane in this experiment?

17. During the 1960s, the concentrations of PCBs in Great Lakes fish started to increase as this pollutant became more and more ubiquitous. The following are PCB concentrations in Lake Michigan bloaters as a function of time (in years, concentration in ppm): 1958, 0.03; 1960, 2.32; 1962, 4.37; 1964, 5.51; 1966, 6.60; 1968, 7.03; and 1972, 7.26. What is the rate constant associated with the accumulation of PCBs in these fish?

18. Let us assume that the average person now has a daily intake of 0.8 pg of dioxin per kg of body weight and that the average concentration of dioxin in that person is 0.7 ppt. What is dioxin's average half-life in people?

19. The Victoria River flows from Lake Albert into Lake George. Water flows into Lake Albert at a rate 25,000 L/s; it evaporates from Lake Albert at a rate 1900 L/s and from Lake George at a rate 2100 L/s. The Elizabeth River flows into the Victoria River between the two lakes at a rate 11,000 L/s. Evaporation from the rivers can be ignored. A soluble pollutant flows into Lake Albert at a rate of 2 mg/s. There are no other sources of the pollutant, it is well mixed in both lakes, and it does not evaporate. Both lakes are at hydrological steady states. In the steady state, what is the concentration of the pollutant in each lake in ng/L?

20. The atmospheric concentration of 2,4,4'-trichloro-biphenyl (a PCB) averages 1.9 pg/m^3 throughout the Earth's atmosphere, and the second-order rate constant for the reaction of this PCB with OH is 1.1×10^{-12} cm^3/s. [P.N. Anderson and R.A. Hites, *Environmental Science and Technology* **30**, 1756–1763 (1996)]. What is the residence time of this PCB in the troposphere due to this reaction? You may assume that the atmospheric concentration of OH is 9.4×10^5 cm^{-3} and that the molecular weight of this PCB is 256.

CHAPTER 3

ATMOSPHERIC CHEMISTRY

Atmospheric chemistry is a vast subject, and it was one of the first areas of environmental chemistry to be developed with some scientific rigor. Part of the motivation for this field was early problems with smog in Los Angeles and with stratospheric ozone depletion. This chapter presents only a quick survey of some of these areas; for more details, one should consult the excellent textbooks by Seinfeld and Pandis[1] or by Finlayson-Pitts and Pitts.[2]

3.1 LIGHT

Light plays an important role in atmospheric chemistry. The energy of the light in the Earth's atmosphere is often needed to initiate reactions. For example, without the energy of light coming from the Sun, there would be no stratospheric ozone. The critical feature of light as a reactant is its wavelength. Thus, a quick review of the electromagnetic spectrum is on order.

---

[1] J.H. Seinfeld and S.N. Pandis, *Atmospheric Chemistry and Physics*, John Wiley & Sons, Inc. New York, 1998, 1326 pp.

[2] B. Finlayson-Pitts and J.N. Pitts, *Atmospheric Chemistry: Fundamentals and Experimental Techniques*, John Wiley & Sons, Inc. New York, 1986, 1098 pp.

Elements of Environmental Chemistry, by Ronald A. Hites
Copyright © 2007 John Wiley & Sons, Inc.

There are several regions of the electromagnetic spectrum; in order of decreasing energy they are as follows:

Region	Wevelengths	Transitions
X rays	0.01–100 Å[3]	inner electrons
Ultraviolet (UV)	100–400 nm	valance electrons
Visible	400–700 nm	valance electrons
Infrared (IR)	2.5–50 μm	molecular vibrations
Microwave	0.1–100 cm	molecular rotations

The relationship of energy to wavelength is

$$E = h\nu = \frac{hc}{\lambda}$$

where

E = energy in Joules (J)
h = Planck's constant = 6.63×10^{-34} J s/molecule
ν = frequency of the light in s^{-1} = c/λ
c = speed of light = 3×10^8 m/s
λ = wavelength of light in meters (m)

We can use this equation to convert from wavelength (or frequency) to energy.

The energy of the oxygen-to-oxygen bond in O_2 is 4.92×10^5 J/mol. What is the maximum wavelength of light that could break this bond?

Strategy. We can rearrange the above equation in terms of wavelength and use the given information

---

[3] Named after the famous Swedish scientist A.J. Ångström; 1 Å = 10^{-8} cm.

$$\lambda = \frac{hc}{E} = \left(\frac{6.63 \times 10^{-34}\,\text{J s}}{\text{molecule}}\right)\left(\frac{3 \times 10^8\,\text{m}}{\text{s}}\right)$$

$$\times \left(\frac{\text{mol}}{4.92 \times 10^5\,\text{J}}\right)\left(\frac{6.02 \times 10^{23}\,\text{molecules}}{\text{mol}}\right)$$

$$\times \left(\frac{10^9\,\text{nm}}{\text{m}}\right)$$

$$= 243\,\text{nm or less}$$

This wavelength is in the UV region, and it does not reach the surface of the Earth.

3.2 ATMOSPHERIC STRUCTURE

The temperature and pressure of the Earth's atmosphere change as a function of altitude; see Figure 3.1. From the Earth's surface to about 15 km in height, the temperature drops at about 6.5 deg/km; this is called the lapse rate. At about 10–15 km, the temperature starts to increase. This altitude, where this rate changes, is called the tropopause. The zone between the surface and the tropopause is called the troposphere. The temperature increase continues from the tropopause up to about 50 km, and this altitude is called the stratopause. The zone of the atmosphere between the tropopause and the stratopause is called the stratosphere. The pressure decreases exponentially throughout the atmosphere's height (see Fig. 3.1). The pressure at any given altitude (z, in km) is given by

$$P_z = P_0 e^{-z/7}$$

where P_0 is the pressure at sea level (1 atm or 760 Torr).

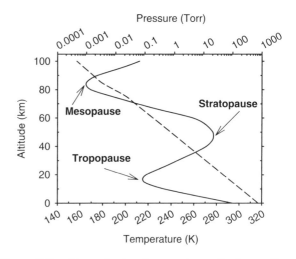

Figure 3.1 Atmospheric temperature (bottom, linear scale, solid line) and atmospheric pressure (top, logarithmic scale, dashed line) as a function of altitude. The average altitudes dividing the troposphere from the stratosphere (the tropopause), the stratosphere from the mesosphere (the stratopause), and the mesosphere from the thermosphere (the mesopause) are also shown. [Replotted from B.J. Finlayson-Pitts and J.N. Pitts, p. 9.]

Because of the rotation of the Earth and because of the warm equatorial and cool polar regions, the atmosphere divides itself into six regions (roughly corresponding to the climatic zones). These atmospheric regions are called the northern and southern Polar, Ferrel, and Hadley cells (so called because they are "roll cells"); see Figure 3.2. This division of the atmosphere tends to slow the mixing

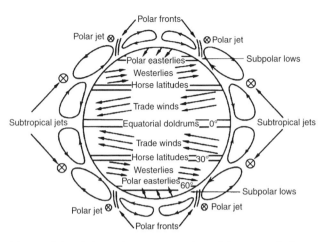

Figure 3.2 Schematic representation of the general circulation of the Earth's atmosphere. [Reproduced from Seinfeld and Pandis, p. 12; used by permission].

of pollutants emitted into one cell from entering the other cells. For example, it takes many years for pollutants emitted into the Northern Hemisphere, where most people live, to mix into the Southern Hemisphere.

3.3   OZONE

3.3.1 Introduction to Ozone

Ozone (O_3) is a light blue gas (boiling point $= -110°C$). It has a unique "electric" odor and is highly reactive. Its structure is three oxygen atoms linked together in the shape of a V (it is not cyclic); the O—O—O bond angle is 127°.

Ozone absorbs UV light in the range of 200–300 nm, and it has a maximum concentration of about 10 ppm in the atmosphere at about 35 km height (in the stratosphere). Thus, ozone acts as the Earth's UV shield (or sunscreen), preventing UV radiation damage to the biosphere. In effect, ozone prevents light of wavelengths less than 300 nm from reaching the Earth's surface. If we lose the ozone sunscreen, we might have an increase in skin cancer and eye cataracts and a decrease in photosynthesis. Lower stratospheric temperatures could also result from ozone depletion.

The main set of reactions that describe the production and loss of ozone in the stratosphere are the Chapman reactions:

$$O_2 + h\nu \rightarrow 2\ O \qquad \text{wavelengths} < 240\ nm$$
$$O + O_2 + M \rightarrow O_3 + M^* \quad \text{M is any third molecule}$$
$$O_3 + h\nu \rightarrow O_2 + O \qquad \text{wavelengths} < 325\ nm$$
$$O + O_3 \rightarrow 2\ O_2$$

In this case, M^* represents a vibrationally excited oxygen or nitrogen molecule. As this vibrational energy dissipates, the oxygen and nitrogen molecules tend to move faster, which is perceived by the observer (us) as heat. Thus, the stratosphere is warmer than the troposphere. Note that the stratosphere is the only region of the Earth's atmosphere where there is both enough UV radiation and enough gas pressure such that the molecules and atoms can react and collide. The Chapman reactions account for most, but not all, of the ozone reactions in the stratosphere—more on this later.

Other important ozone reactions are all based on the general catalytic cycle:

$$X + O_3 \rightarrow XO + O_2$$
$$\underline{XO + O \rightarrow O_2 + X}$$
$$O + O_3 \rightarrow 2\ O_2 \qquad \text{net reaction}$$

Note that the third reaction is the total (or net) reaction of the first two reactions and that X does not form or disappear in this overall reaction. Thus, X is a catalyst for this reaction.

3.3.2 Ozone Catalytic Cycles

There are three important catalytic cycles or three ways to lose your ozone:[4]

a. NO/NO_2
b. OH/OOH
c. Cl/OCl

---

[4] With apologies to Paul Simon's song "Fifty Ways to Leave Your Lover."

The NO/NO$_2$ pathway. One source of NO is the reaction of N$_2$O with excited oxygen atoms O:

$$N_2O + O \rightarrow 2\ NO$$

In this case, N$_2$O (called nitrous oxide or laughing gas) has natural sources, such as emissions from swamps and other oxygen-free ("anoxic") waters and soils. The oxygen atoms in this reaction can come from several tropospheric photolytic reactions involving OH or OOH. Another source of NO is the thermal reaction between N$_2$ and O$_2$:

$$N_2 + O_2 \rightarrow 2\ NO$$

This reaction requires very high temperatures, and thus, it occurs mostly in combustion systems such as automobile and jet engines. This reaction is also likely to occur in a thermonuclear explosion, although the production of NO from such an event would be the least of our problems.

Once NO has formed, it is destroyed by reactions with ozone according to the following coupled reactions:

$$NO + O_3 \rightarrow NO_2 + O_2$$
$$\underline{NO_2 + O \rightarrow O_2 + NO}$$
$$O + O_3 \rightarrow 2\ O_2 \qquad \text{net reaction}$$

Notice that the net result is the loss of ozone. This reaction is one of the reasons we do not have a fleet of supersonic airplanes exhausting NO into the lower stratosphere.

The OH/OOH pathway. First a bit of terminology: OH is called the hydroxyl radical and OOH is called the hydroperoxyl radical. There are several ways to form OH in the atmosphere. One such reaction is the reaction of methane with excited atomic oxygen:

$$CH_4 + O \rightarrow OH + CH_3$$

In this case, the methane comes from natural sources such as anoxic waters and soils and from cow flatus. Once formed, OH reacts with ozone to form OOH, which is in turn lost by reactions with atomic oxygen.

$$OH + O_3 \rightarrow OOH + O_2$$
$$\underline{OOH + O \rightarrow OH + O_2}$$
$$O + O_3 \rightarrow 2\ O_2 \qquad \text{net reaction}$$

The Cl/OCl pathway. This process needs atomic chlorine to get started. The chlorine atoms react with ozone to produce ClO, which is in turn lost by reactions with atomic oxygen for a net loss of ozone:

$$Cl + O_3 \rightarrow ClO + O_2$$
$$\underline{ClO + O \rightarrow Cl + O_2}$$
$$O + O_3 \rightarrow 2\ O_2 \qquad \text{net reaction}$$

The importance of this process was not recognized until the mid-1970s when it was realized that Cl could be produced from the photodegradation of chlorofluorocarbons (CFCs), which had become very widely used as refrigerants and for other applications for which an inert gas was needed. This realization was a big deal, and it eventually earned Mario Molina, F. Sherwood Rowland, and Paul Crutzen the Nobel Prize for chemistry in 1995.

Let us digress for a moment and discuss the history of CFCs. Refrigerants are gases that can be compressed and expanded, and as they do so they move heat from one place to another. For example, a refrigerator[5] uses a gas and a compressor to move heat from inside the box to outside the box. In the early part of the twentieth century, gases such as NH_3 and SO_2 were used as refrigerants. Unfortunately, these gases were toxic and reactive, and people were actually killed by their leaking refrigerators. In 1935, DuPont invented Freon 11 (CCl_3F) and Freon 12 (CCl_2F_2), and these compounds turned out to be almost perfect refrigerants. The Freons were chemically stable, nonflammable, and nontoxic. Eventually, the worldwide production of these compounds reached about 10^9 kg/year. By the 1970s, it was found that these CFCs were so stable that they had no sinks in the troposphere. They were neither water soluble nor reactive with OH (or anything else). As a result, their tropospheric residence times were on the order of 100 years.

---

[5] As George Carlin says, "This is a device to frigerate again."

In 1974, Rowland and Molina[6] suggested that the only environmental sink for these compounds was transport into the stratosphere where the CFCs would be photolyzed to Cl, which would in turn react with ozone. For example, Freon 12 will photolyze in the stratosphere to produce chlorine atoms:

$$CCl_2F_2 + h\nu \rightarrow CF_2Cl + Cl \ (\lambda < 250 \, nm)$$

The Cl produced from these reactions enters into the cycle shown above.

The atmospheric concentrations of CFCs had been going up rapidly, doubling every 10 years or so, and by the early 1980s, many people were concerned about the effect of CFCs on stratospheric ozone, but many were not convinced that this effect was real. Nevertheless, by the late 1970s, nonessential uses were restricted (e.g., aerosol can propellants). The true proof of the concept of anthropogenic ozone depletion occurred in 1985, when the Antarctic ozone hole was discovered. The Antarctic hole grows each winter (August in the Antarctic) and then collapses each spring. Since the discovery of the ozone hole, the amount of ozone destroyed and the size of the hole has generally increased each year.

Why in Antarctica? A minor reason is that the southern polar atmosphere is well contained during the

---

[6] M.J. Molina and F.S. Rowland, Stratospheric sink for chlorofluorocarbons. Chlorine atom-catalyzed destruction of ozone. *Nature*, **249**, 810–812 (1974).

winter compared to the Arctic and thus only slowly mixes with the air in the rest of the Southern Hemisphere. As a result, ozone is not easily replenished during the polar winter. However, the most important reason is a series of reactions that take place on the surface of ice particles in the dark. To explain this, we must digress again.

If there were no new inputs of chlorine into the stratosphere, eventually all of the chlorine would be inactivated (i.e., it would eventually not be in the form of Cl_2, Cl, or ClO). The inactivation reactions produce HCl and $ClONO_2$. In the relatively warm months in the Antarctic stratosphere, these two compounds are in the gas phase as opposed to condensed on solid phases (such as ice particles). Note that we will use abbreviations in the reactions to indicate the phases: (s) and (g) mean in the solid and gas phases, respectively.

When it gets cold, small ice crystals form in the Antarctic stratosphere; these are called polar stratospheric clouds. HCl condenses onto these surfaces. Unfortunately for the ozone, there is a reaction of HCl (s) with $ClONO_2$ (g), which is catalyzed by the surface.

$$HCl\,(s) + ClONO_2(g) \rightarrow Cl_2(g) + HNO_3(s)$$

This reaction produces Cl_2 gas, which can form active Cl atoms as soon as photons are available.

$$Cl_2(g) + h\nu \rightarrow 2\,Cl$$

This happens in the early spring in the Antarctic or in the months of September and October. The Cl atoms then catalytically destroy ozone, resulting in the rapid loss of ozone that we call the ozone hole. As time goes by, the Antarctic stratosphere warms up, the ice crystals melt, Cl starts to be inactivated, and the ozone hole heals.

The overall loss of ozone from the stratosphere is now about 30%, and the stratospheric concentration of chlorine is still increasing. Small ozone losses are even being noticed in equatorial regions. The issue of CFCs and ozone depletion is now clear—chlorofluorocarbons cause ozone depletion. As a result of this consensus, industry (at least in the developed world) has largely quit the business; DuPont, for example, quit the business in 1988. The Montreal Protocol now restricts the manufacture and sale of CFCs on a global basis. This protocol has been revised twice to make it more stringent, and it required a complete global ban on these compounds in industrialized countries by 1996. Hydrogenated CFCs such as $CHCl_2F$, which should degrade in the troposphere, are replacing the old CFCs. But even these compounds will be banned (in the United States at least) by the year 2020.

3.4 CHEMICAL KINETICS

3.4.1 Pseudo-Steady-State Example

Many chemical reactions involve very reactive intermediate species such as free radicals, which as a result of their high reactivity are consumed virtually as fast as they are formed and consequently exist at very low concentrations. The pseudo-steady-state approximation is a fundamental way of dealing with these reactive intermediates when deriving the overall rate of a chemical reaction mechanism.

It is perhaps easiest to explain the pseudo-steady-state approximation by way of an example. Consider the simple reaction $A \rightarrow B + C$, whose elementary steps consist of the activation of A by collision with a background molecule M (in the atmosphere M is typically N_2 and O_2) to produce an energetic A molecule denoted by A^*, followed by the decomposition of A^* to give B and C. Thus, we write the mechanism as

$$A + M \rightarrow A^* + M \quad k_1$$
$$A^* + M \rightarrow A + M \quad k_{-1}$$
$$A^* \rightarrow B + C \qquad k_2$$

Note that each reaction has a rate constant and that the second reaction is the reverse of the first (i.e., A^* may be deactivated by collision with M). Assuming the Earth's atmosphere as one big compartment, we have one way of losing A and one way of forming A; hence, its rate of change in the atmosphere is

$$\frac{d[A]}{dt} = -k_1[A][M] + k_{-1}[A^*][M]$$

In this notation, the square brackets mean the concentration of the atom or compound in question; for example, $[O_2]$ is 10^{17} molecules/cm^3 at 30 km altitude.

In this equation, the derivative is positive and means the rate at which A is formed. The first term on the right is negative because A is being lost by the first reaction. The second term is positive because A is being formed by the second reaction. Both reactions are of second order,[7] so both terms have two concentrations and a rate constant.

The rate of formation of A^* is given by

$$\frac{d[A^*]}{dt} = k_1[A][M] - k_{-1}[A^*][M] - k_2[A^*]$$

There are three terms here because there are two ways to lose A^* and one way to form it. The reactive intermediate in this system of reactions is A^*. The pseudo-steady-state approximation states that the rate of formation of A^* is equal to its rate of loss; in other words, $[A^*]$ does not change over time. Thus,

$$\frac{d[A^*]}{dt} = 0$$

This in turn says that

$$0 = k_1[A][M] - k_{-1}[A^*][M] - k_2[A^*]$$

This expression can now be solved for $[A^*]$ and that result substituted back into the expression for the rate of

---

[7] Second order means that two atoms or molecules have to collide so that the reaction will happen. Obviously, if the concentration of one or both of the reactants is low, the reaction will be slow. Thus, the rate of the reaction is proportional to the concentration of both reactants times each other, and the proportionality constant is the rate constant k.

formation of A to get

$$\frac{d[A]}{dt} = -\frac{k_1 k_2 [A][M]}{k_{-1}[M] + k_2}$$

Before we go any further, please do the algebra to make sure this expression is correct.

This expression tells us that the rate of formation of A depends not only on the concentration of A, but also on the concentration of M, which is proportional to the total pressure of the atmosphere at the height where the reactions are taking place. Notice that if $k_{-1}[M] \gg k_2$, then

$$\frac{d[A]}{dt} = -\frac{k_1 k_2 [A][M]}{k_{-1}[M]} = -\frac{k_1 k_2}{k_{-1}}[A] = -k'[A]$$

This means that the reaction rate is of first order, depending only on [A]. If $k_{-1}[M] \ll k_2$, then

$$\frac{d[A]}{dt} = -\frac{k_1 k_2 [A][M]}{k_2} = -k_1 [A][M]$$

In this case, the reaction is of second order, depending on both [A] and [M], and the rate depends only on k_1. This is an example of a situation in which the rate of one reaction in a reaction sequence (reaction 1 in this case) controls the rate of the whole sequence; this is known in the trade as a rate-determining step or reaction.

3.4.2 Arrhenius Equation

Rates of reactions change as a function of the temperature of the reactants. Frequently, reactions go faster when the temperature is higher—this is the basis of cooking food. The Arrhenius equation relates the rate constant of a reaction to the temperature at which the reaction is taking place. In its simplest form, the Arrhenius[8] equation is

$$k = A \exp\left(\frac{-E_A}{RT}\right)$$

where A is the so-called "pre-exponential factor" (which has the same units as the rate constant), E_A is the activation energy of the reaction (in units of kJ/mol), T is the temperature (in deg K), and R is the gas constant (in this case, 8.3 J/deg mol). We can and will use this equation to calculate the rate constants of various reactions at different temperatures in the atmosphere.

3.4.3 Chapman Reaction Kinetics

As you know, the Chapman equations are

$$O_2 + h\nu \rightarrow 2\,O \qquad\qquad k_1 = 10^{-11.00}\,s^{-1}$$
$$O + O_2 + M \rightarrow O_3 + M^* \qquad k_2 = 10^{-32.97}\,cm^6/s$$
$$O_3 + h\nu \rightarrow O_2 + O \qquad\qquad k_3 = 10^{-3.00}\,s^{-1}$$
$$O + O_3 \rightarrow 2\,O_2 \qquad\qquad k_4 = 10^{-14.94}\,cm^3/s$$

Notice that we have now numbered these equations from 1 to 4 and supplied the rate constants for these

---

[8] Incidentally, Professor Hites was forced to have lunch in Arrhenius' house on the campus of Stockholm University as penitence for thinking that Arrhenius was Danish—he was Swedish.

equations at an altitude of 30 km, where $T = 233$ K and $P = 0.015$ atm. I am violating our normal rule about too many significant figures here because we will eventually need all four such figures to solve this problem.

What is the concentration of ozone at 30 km altitude?

Strategy. Let us assume that the Earth's stratosphere is a large homogeneous compartment and that the flows of O_2, O, and O_3 are given by the four Chapman equations. The concentration of M (N_2 and O_2) is sufficiently high so that it is virtually a constant. To solve this problem, let us first set up the equations for the steady-state concentrations of O and O_3; in other words, we will set up the equations for the rates of formation of O and O_3 and set these rates equal to zero. Using these two expressions, we will then calculate the value of the O_3 to O_2 ratio at 30 km and from this ratio get $[O_3]$.

Remember from our previous work that at 30 km $[O_2]$ is $10^{17.00}$ cm^{-3}, and therefore, $[M] = 10^{17.00}/0.21 = 10^{17.68}$ molecules/cm^3. From the four Chapman equations, we see that the equations for the formation of O_3 and O are

$$\frac{d[O_3]}{dt} = k_2[O_2][O][M] - k_3[O_3] - k_4[O][O_3]$$

$$\frac{d[O]}{dt} = 2k_1[O_2] - k_2[O_2][O][M] + k_3[O_3] - k_4[O][O_3]$$

Because the concentrations of both O_3 and O are not changing much as a function of time in the atmosphere, we can set both of these rates equal to zero (the pseudo-steady-state approximation). Now we have two equations with two unknowns (O and O_3). The easiest way to solve these equations is to first subtract the second from the first and get

$$-2k_1[O_2] + 2k_2[O_2][O][M] - 2k_3[O_3] = 0$$

Let us assume that $k_1[O_2] \ll k_3[O_3]$; in other words, we assume that

$$\frac{k_1}{k_3} \ll \frac{[O_3]}{[O_2]}$$

which seems fair given that k_1/k_3 is about 10^{-8}, but we will check this later. Thus, the above equation is simplified to

$$k_2[O_2][O][M] = k_3[O_3]$$

Hence,

$$[O] = \frac{k_3[O_3]}{k_2[O_2][M]}$$

Now we have to go back and add the two steady-state equations for the formation of O_3 and O and get

$$2k_1[O_2] - 2k_4[O][O_3] = 0$$

Into this equation, we substitute the expression for [O] we just found above and get

$$2k_1[O_2] - \frac{2k_3k_4[O_3]^2}{k_2[O_2][M]} = 0$$

We can rearrange this to

$$\frac{[O_3]^2}{[O_2]^2} = \frac{k_1 k_2 [M]}{k_3 k_4}$$

Hence,

$$[O_3] = \left(\frac{k_1 k_2 [M]}{k_3 k_4}\right)^{\frac{1}{2}} [O_2] = \left(\frac{10^{-11.00} 10^{-32.97} 10^{17.68}}{10^{-3.00} 10^{-14.94}}\right)^{0.5}$$
$$\times 10^{17.00}$$
$$= 10^{12.83} \approx 10^{13} \text{ cm}^{-3}$$

Now would be a good time to verify that the result of this calculation is correct and that the $k_1/k_3 \ll [O_3]/[O_2]$ assumption is acceptable.

The problem with this result is that it is wrong. This result is quite a bit too high. The error is because we omitted the other ways to lose your ozone, the most important of which is the NO/NO$_2$ cycle. Let us add these two equations and rate constants (at 30 km) and calculate how much (in ppb) of these gases it takes to reduce the O$_3$ by a factor of 3.

$$NO + O_3 \rightarrow NO_2 + O_2 \qquad k_5 = 10^{-14.31}$$
$$NO_2 + O \rightarrow NO + O_2 \qquad k_6 = 10^{-10.96}$$

Now we need three steady-state equations, one for O$_3$, one for O, and one for NO$_2$:

$$\frac{d[O_3]}{dt} = k_2 [O_2][O][M] - k_3 [O_3] - k_4 [O][O_3]$$
$$- k_5 [NO][O_3]$$

$$\frac{d[O]}{dt} = 2k_1[O_2] - k_2[O_2][O][M] + k_3[O_3] - k_4[O][O_3]$$
$$\qquad\qquad - k_6[NO_2][O]$$
$$\frac{d[NO_2]}{dt} = k_5[NO][O_3] - k_6[NO_2][O]$$

We can set all three of these equations equal to zero (for the steady-state assumption). If we subtract the second from the first and add the third equation and if we make the same assumption about $k_1[O_2]$ being very small, we get the same expression for [O], namely

$$[O] = \frac{k_3[O_3]}{k_2[O_2][M]}$$

Now if we simply set all three of these equations equal to zero and add them all together, we get

$$2k_1[O_2] - 2k_4[O][O_3] - 2k_6[NO_2][O] = 0$$

Canceling the factor of 2 and substituting the expression of [O] from just above, we get

$$k_1[O_2] = \frac{k_3k_4[O_3]^2}{k_2[O_2][M]} + \frac{k_3k_6[O_3][NO_2]}{k_2[O_2][M]}$$

The measured concentration of NO_2 is about 7 ppb, which converts to a number density of $10^{9.53}$ cm^{-3}. We can substitute this into the above equation using the given rate constants and get

$$10^{-11.00}10^{17.00} = \frac{10^{-3.00}10^{-14.94}[O_3]^2}{10^{-32.97}10^{17.00}10^{17.68}}$$
$$+ \frac{10^{-3.00}10^{-10.96}10^{9.53}[O_3]}{10^{-32.97}10^{17.00}10^{17.68}}$$
$$10^{6.00} = 10^{-19.65}[O_3]^2 + 10^{-6.14}[O_3]$$

This is a quadratic equation in terms of the ozone concentration. When faced with this level of complexity, it is often convenient to guess the answer and to see if one of the three terms can be dropped because it is small relative to the others. We know that the correct ozone concentration is quite a bit lower than 10^{13} cm^{-3}; let us guess that it is 10^{12} cm^{-3}. Using this value, the three terms in the quadratic equation become

$$10^6 = 10^{4.3} + 10^{5.9}$$

Thus, the first term on the right is about a factor of 40 smaller than the others, and we can drop it and see if things work out. Thus,

$$[O_3] = \frac{10^{6.00}10^{-32.97}10^{17.00}10^{17.68}}{10^{-3.00}10^{-10.96}10^{9.53}}$$
$$= 10^{12.14} = 1.4 \times 10^{12} \text{ cm}^{-3}$$

This result is not too far from our guess, so we were justified in dropping that term.

We can also use the quadratic equation and solve for the ozone concentration. Remember that for an equation of the type

$$ax^2 + bx + c = 0$$
$$x = \frac{-b \pm \sqrt{b^2 - 4ac}}{2a}$$

Hence,

$$
\begin{aligned}
[O_3] &= \frac{-10^{-6.14} + \sqrt{10^{-12.28} + 4 \times 10^{-19.65}10^{6.00}}}{2 \times 10^{-19.65}} \\
&= \frac{\sqrt{10^{-12.28} + 10^{-13.05}} - 10^{-6.14}}{10^{-19.35}} \\
&= \frac{\sqrt{10^{-12.21}} - 10^{-6.14}}{10^{-19.35}} = \frac{10^{-6.11} - 10^{-6.14}}{10^{-19.35}} \\
&= \frac{10^{-7.29}}{10^{-19.35}} = 10^{12.1} \text{ cm}^{-3}
\end{aligned}
$$

Notice that we needed all of the four significant figures to make this work out. This ozone concentration is the same as the value we got from the simpler calculation above, as it should be.

We should not take this result too literally; the actual concentration of ozone in the stratosphere varies a lot depending on light incidence and global location. Nevertheless, these calculations demonstrate the necessity of including the NO_x reactions in any calculation of stratospheric ozone concentrations. This approach also demonstrates the power of a kinetic modeling approach —once you know the mechanism of the reactions and the rate constants (preferably as a function of temperature), you can figure out the most amazing things.

3.5 SMOG

There are two kinds of smog (= smoke + fog): (a) Reducing smog is largely based on SO_2 and was prevalent in London, England, in the 1950s. This has mostly disappeared due to emission regulations. (b) Oxidizing smog is also called photochemical smog, and this is a big problem in Los Angeles, California, and in many cities in the United States and around the world (e.g., in Tokyo, Paris, and Mexico City). Both types of smog cause eye irritation and lung damage; they can also have severe agricultural effects. We will focus only on photochemical smog.

To make photochemical smog, we need four things:

1. warm air (hotter than about 290 K = 63°F);
2. lots of intense sunlight ($h\nu$);
3. lots of hydrocarbons and NO_x (which usually means lots of cars);
4. stable air masses (e.g., a city surrounded by mountains).

Los Angeles meets these requirements.

Photochemical production of ozone in the troposphere occurs from the photolysis of nitrogen dioxide (NO_2) during the daytime, producing oxygen atoms (O):

$$NO_2 + h\nu \rightarrow NO + O \qquad (a)$$

In the troposphere, the oxygen atoms react quickly with oxygen molecules (O_2), producing ozone (O_3):

$$O + O_2 \rightarrow O_3 \qquad (b)$$

Ozone can also react with nitric oxide, producing NO_2:

$$O_3 + NO \rightarrow NO_2 + O_2 \qquad (c)$$

These three reactions result in a fast cycle where ozone is produced through reactions (a) and (b), but destroyed through reaction (c). As a result, concentrations of ozone quickly reach a steady-state concentration during the daytime, with no net production of ozone.

There are several steps for making smog. They happen in sequence during the day.

Step 1. Cars generate NO thermally.

$$N_2 + O_2 \rightarrow 2\,NO$$

These NO emissions reduce the steady-state concentration of ozone due to reaction (c). However, cars also emit carbon monoxide and a variety of hydrocarbons (HC) as a result of incomplete combustion. These emissions react with the hydroxyl radical to produce peroxy radicals:

$$OH + CO\,(+O_2) \rightarrow HO_2 + CO_2 \qquad (d)$$
$$OH + HC\,(+O_2) \rightarrow RO_2 + H_2O \qquad (e)$$

where R represents any organic radical such as CH_3 or C_2H_5. Not all hydrocarbons are equally reactive. For example, methane and acetylene are not particularly reactive, but toluene, propylene, pinene, and isoprene are very reactive. The latter are often called "non-methane hydrocarbons" for short.

Step 2. The peroxy radicals produced from the above reaction oxidize the NO emissions to NO_2 (which gives smog its characteristic yellow color).

$$HO_2 + NO \rightarrow OH + NO_2 \tag{f}$$

$$RO_2 + NO\ (+O_2) \rightarrow R'O + NO_2 + HO_2 \tag{g}$$

Step 3. The conversion of NO emissions from cars to NO_2 by the reactions of CO and hydrocarbons increases the concentration of ozone during the day. NO_2 and hydrocarbons react to give aldehydes (e.g., acetaldehyde, CH_3CHO), ozone, and peroxyacetyl nitrate (PAN). A typical hydrocarbon is ethane (CH_3CH_3). Thus,

$$CH_3CH_3 + OH\ (+O_2) \rightarrow CH_3CH_2O_2 + H_2O$$

$$CH_3CH_2O_2 + NO\ (+O_2) \rightarrow CH_3CHO + HO_2 + NO_2$$

$$OH + CH_3CHO \rightarrow CH_3CO^\bullet + H_2O$$

$$CH_3CO^\bullet + O_2 \rightarrow CH_3COOO^\bullet$$

$$CH_3COOO^\bullet + NO_2 \rightarrow CH_3COOONO_2$$

This last compound is PAN, which is an eye irritant and has the following structure:

To eliminate smog in problem cities, one could cut back on the NO and hydrocarbon emissions. This could mean cutting back on the number of cars, which is usually not politically feasible, especially in California. The alternate is to cut back on the emissions from each car, and the automobile industry has done this to a considerable extent, at least in the United States. For example, the hydrocarbon emission rate has changed from about 9 g of hydrocarbons per mile driven in 1968 to well below 0.4 g/mile now. These reductions are the result of a relatively modern device in cars called the two-stage catalytic converter. The first stage uses ruthenium to reduce NO to N_2, and the second stage uses platinum and/or palladium to oxidize hydrocarbons to CO_2.

3.6   GREENHOUSE EFFECT

To understand the "greenhouse effect," we first need to understand the emission of light from an object (such as the Earth) as a function of the temperature of that object. The spectrum of light (remember that a spectrum is the light intensity as a function of wavelength or frequency) from an object that does not create any light of its own is called the spectrum of a "blackbody." The equation of this spectrum (for a given temperature) was worked out long ago by Max Planck, and not surprisingly it is called "Planck's law":

$$E(\lambda) = \frac{2\pi hc^2 \lambda^{-5}}{\exp\left(\frac{hc}{\lambda kT}\right) - 1}$$

where

$E(\lambda)$ = energy at wavelength λ (in units of W/m^3); remember that a Watt (W) is a Joule per second

h = the Planck constant (6.63×10^{-34} J s/molecule)

c = speed of light (3×10^8 m/s)

λ = wavelength of light (in meters)

$k =$ the Boltzmann[9] constant $(1.38 \times 10^{-23} \, \text{J/deg})$
$T =$ temperature (K)

There are two features of this function that are good to know. The first is the maximum wavelength for a given temperature:

$$\lambda_{max} = \frac{2900 \, \mu\text{m deg}}{T}$$

where the wavelength is in microns (μm). This equation is called Wein's law. The clever student can derive this result from Planck's law (see the problem set).

The second feature of Planck's law that is good to know is the total energy emitted by a blackbody at a given temperature. This is the integral of this equation with respect to wavelength from 0 to ∞. One gets

$$E_{\text{total}} = \frac{2\pi^5 k^4 T^4}{15 c^2 h^3} = 5.67 \times 10^{-8} T^4 = \sigma T^4$$

where E_{total} is in W/m^2 and $\sigma = 5.67 \times 10^{-8} \, \text{W/m}^2\text{deg}^4$. The latter is called the Stefan – Boltzmann constant. Just for fun, please verify for yourself that the Stephan–Boltzmann constant (σ) is calculated correctly; be sure to keep track of the units.

---

[9] Although Boltzmann was from southern Germany, there is no truth to the rumor that his given name was Billy-Bob (although the alliteration is pleasing). His first name was actually Ludwig.

What are the maximum wavelengths and total energies emitted by the Sun and by the Earth?

Strategy. Let us assume that the Sun has a surface temperature (T) of 6000 K. Thus,

$\lambda_{max} = 2900/6000 = 0.48\,\mu m = 480\,nm$ (which is in the visible region)
$E_{total} = 5.67 \times 10^{-8} \times 6000^4 = 7.35 \times 10^7\,W/m^2$

Now let us assume that the Earth has a surface temperature of 288 K. Thus,

$\lambda_{max} = 2900/288 = 10.1\,\mu m$ (which is in the infrared region)
$E_{total} = 5.67 \times 10^{-8} \times 288^4 = 3.90 \times 10^2\,W/m^2$

Note that the Earth emits about 200,000 times less energy than the Sun (good!) and that the Earth emits mostly in the infrared (which is mostly heat).

Now we can use this idea to calculate the temperature of the Earth. We can do this by balancing the energy coming into the Earth against the energy leaving the Earth. The energy coming in is due to the emission of the Sun at the distance of the Earth's orbit. This energy is given by the solar constant, which is usually called Ω, and it is 1372 W/m^2.

This might be a little confusing given our previous calculation of the total energy emitted from a blackbody at 6000 K. Remember that this value of E_{total} of 7.35×10^7 W/m^2 was for the Sun at its surface, but the Earth is 1.5×10^8 km away from the center of the

Sun. Hence, we must diminish this intensity by the inverse square law:

$$I \propto \frac{1}{d^2}$$

where I is the intensity and d is the distance. The radius of the Sun is 6.5×10^5 km; hence, the dilution of light is a factor of

$$\left(\frac{1.5 \times 10^8}{6.5 \times 10^5}\right)^2 = 53{,}300$$

Hence,

$$\Omega = \frac{7.35 \times 10^7}{53{,}300} = 1372 \, \text{W/m}^2$$

This energy is distributed over the Earth's surface, which is roughly a sphere, but if you were at the Sun looking at the Earth (using a well-cooled telescope), all you would see is a disk. Thus, the energy that arrives at the Earth's orbit must be "diluted" by the ratio of the area of a sphere $(4\pi r^2)$ divided by the area of a disk (πr^2), which is exactly a factor of 4. In addition, some of the incoming energy does not make it to the surface of the Earth; it is reflected away by, for example, clouds. The fraction reflected is called the albedo, usually abbreviated a. On average, the Earth's albedo is 30%; that is, 30% of the light coming to the Earth is reflected back to space. That is how the astronauts were able to see the Earth from the Moon.

The output side of the energy balance is just the energy of the Earth as a blackbody, and it is given by σT^4. Thus, the Earth's energy balance is given by

$$\sigma T^4 = \frac{(1-a)\Omega}{4}$$

Given the values of σ, Ω, and a, please calculate the temperature of the Earth.

Strategy. We can just substitute the known values of these three constants in this equation and solve for T. If you cannot take a fourth root, then just take a square root twice.

$$T^4 = \left(\frac{(1-a)\Omega)}{4\sigma}\right) = \left(\frac{(1-0.30) \times 1372}{4 \times 5.67 \times 10^{-8}}\right)$$
$$= 4.235 \times 10^9$$
$$T = \sqrt[4]{4.235 \times 10^9} = 255\,\text{K}$$

Thus, the temperature of the Earth's atmosphere should be $255\,\text{K}$ or $-18°\text{C}$. Given that the correct value is $288\,\text{K}$ or $+15°\text{C}$, our calculation is 33 K too low, which is a lot.[10] What is wrong? The answer is the greenhouse effect. This means that the Earth's atmosphere traps some of the heat in the atmosphere. The incoming light has wavelengths in the visible range; it warms up the surface of the Earth, and light is emitted in the infrared range. Gases such as H_2O, CO_2, N_2O, and CH_4 in the atmosphere are transparent to visible but not to infrared radiation; thus, infrared (heat) is absorbed by the atmosphere. And like a blackbody, the atmosphere emits radiation. Thus, the surface of the Earth receives energy from both the Sun and the atmosphere. Although water

---

[10] Or to quote Carl Sagan, "This is a lot—even for a chemist."

vapor is primarily responsible for the natural green-house effect, CO_2 is usually considered the primary global warming culprit, because the concentration of this gas is increasing as a result of human activity.

Based on this discussion, the correct equation for the Earth's overall energy balance should be

$$\sigma T^4 = \frac{(1-a)\Omega}{4} + \Delta E$$

Remember this!

Using the above equation, please calculate the magnitude of ΔE relative to the other two terms for an atmospheric temperature of 288 K.

Strategy. Given that we know all of these numbers, we can just plug and chug. The three terms are

$$\sigma T^4 = (5.67 \times 10^{-8})(288^4) = 390.08 \text{ W/m}^2$$

$$(1-a)\left(\frac{\Omega}{4}\right) = 0.7\left(\frac{1372}{4}\right) = 240.10 \text{ W/m}^2$$

$$\Delta E = 390.08 - 240.10 = 149.98 \text{ W/m}^2$$

It is obvious that the greenhouse term is not small – in fact, it is ~40% of the blackbody term (σT^4). The exact value of ΔE will change depending on the concentrations of those gases in the atmosphere that absorb infrared radiation.

From this equation, we see that there are three factors that control the temperature of the Earth: the albedo, the solar constant, and greenhouse gas concentrations. Let us look at these in turn:

Albedo. The Earth's average albedo is 30%, but it can vary widely. For example, the albedo of snow is about 80%, and the albedo of a forest is about 15%. Thus, as the Earth warms up, the snow might melt, and the albedo might go down. This would cause the temperature to increase even faster. This is an example of a positive feedback; that is, the *rate* of warming of the Earth's atmosphere might actually increase over time. The albedo is also affected by particles in the atmosphere. The biggest source of particles on a global basis is usually volcanoes. For example, in 1815, Mount Tambora exploded putting about 2×10^{11} t of rock dust into the atmosphere. This caused about a 0.4–0.7 K drop in temperature the following year.[11] Given that both CO_2 and particles are released by combustion systems, it is possible that some increases in temperature due to increasing CO_2 concentrations might be offset by increasing particle concentrations. This is an example of a negative feedback system.

Solar constant. This does not change much (less than $\pm 2\%$). Most of the change is due to sunspots, which change on an 11-year cycle. Generally, this factor is ignored by policy wonks, in part, because we cannot do anything about it.

---

[11] R.B. Stothers, The great Tambora eruption in 1815 and its aftermath. *Science*, **224**, 1191–1198 (1984).

Greenhouse gases. CO_2 is the result of the anthropogenic combustion of fossil fuels taking place all over
the globe. Its concentration is now increasing at about
0.4%/year based on good quality measurements. CH_4
and N_2O are also increasing in concentration at rates of
about 0.6% and 0.2%, respectively. These two gases are
the result of agricultural practices, which are increasing
from population pressures. As the concentrations of all
of these gases increase, the value of ΔE increases and
atmospheric temperatures increase.

Is the greenhouse effect real? Yes! Polar ice is disap-
pearing, and polar bears are having trouble finding
food. The Earth's surface temperature (both as measured on the surface and from satellites) is increasing.[12]
It is now virtually certain that[13]

- the concentrations of greenhouse gases have
  increased as a result of anthropogenic activities,
  and this increases the heat retention of the planet;
- the effects of greenhouse gases can last for many
  centuries;
- the Earth's surface has warmed by about $0.5 \pm 0.2°C$
  over the last 100 years;
- the stratospheric temperature has decreased by about
  $1°C$ because of ozone depletion, and it is likely to
  continue to cool as the lower atmosphere absorbs
  more of the Earth's radiation;

---

[12] R.A. Kerr, No doubt about it, the world is warming. *Science*,
312, 825 (2006).
[13] J.D. Mahlman, Uncertainties in projections of human-caused
climate change. *Science*, **278**, 1416–1417 (1997).

- doubling of the atmospheric CO_2 concentration is likely to lead to a $3.0 \pm 1.5°C$ increase in the atmospheric temperature;
- by 2100, it is likely that sea levels will be up by 50 ± 25 cm.

What can we do about this now? The general solution is to avoid fuels that are burned and produce CO_2 as a result of the combustion process, but this is very difficult to do—we are all addicted to our cars and sports utility vehicles and to warm (and cool) homes. It is interesting to note that nuclear power does not produce greenhouse gases, but this does not seem to be a political option, at least in the United States.

3.7   PROBLEM SET

1. What is the energy of a mole of blue photons? Assume the wavelength is 450 nm. Calculate the same information for infrared light at 3 μm and for ultraviolet light at 250 nm.
2. By how much would the Earth's atmospheric temperature change if the greenhouse effect, the albedo, and the solar constant are *all* increased by 1.5%, relative? Please give your answer to two significant figures.
3. What would be the Earth's atmospheric temperature if the greenhouse effect is increased by 10%?
4. The fires from a major nuclear war could result in so much soot that the Earth's albedo could go up by 20%, relative. What atmospheric temperature change would this cause?

5. Lake James is well mixed, and it has a volume of 10^8 m^3. A single river, flowing at 5×10^5 m^3/day, feeds it. Water exits Lake James through the Henry River; evaporation is negligible. A factory on Lake James claims that it is dumping into the lake less than 25 kg/day of tetrachlorobarnene that it is permitted by law. This factory is the only source of tetrachlorobarnene to this lake; this compound is chemically stable and highly water soluble. The factory manager has refused your request to monitor the effluent discharge from the factory, so you take a sample from the lake and measure a tetrachlorobarnene concentration of 100 µg/L. Was the factory manager telling the truth? Justify your answer quantitatively.

6. Assume that the relationship between atmospheric CO_2 concentration and ΔE is given by

$$\Delta E = 133.26 + 0.044[CO_2]$$

where $[CO_2]$ is the atmospheric concentration of CO_2 in ppm. Let us use this relationship to investigate the interactive aspects of increasing CO_2 levels and increasing albedo. If the ambient atmospheric CO_2 concentration and the albedo were both increasing at a rate of 0.2% per year, what would the Earth's average temperature be in 100 years?

7. What change in albedo resulted from the Mount Tambora eruption? The average temperature in the Northern Hemisphere dropped by 0.6°C in 1816. [R.B. Stothers, *Science*, **224**, 1191–1198 (1984).]

8. Nitrous oxide (N_2O) is produced by bacteria in the natural denitrification process. It is chemically inert in the troposphere, but in the stratosphere it is degraded photochemically. The average concentration of N_2O in the troposphere is about 300 ppb, and its residence time there is 10 years. What is the global rate of production of N_2O in units of kg/year? Assume that the volume of the stratosphere (at 0°C and 1 atm) is 10% that of the atmosphere.

9. The constant in Wein's law is usually given as 2900 μm deg. Derive an equation by which this constant can be calculated. *Hint*: Take the derivative of Planck's law with respect to wavelength and set it equal to zero.

10. A Dobson unit (DU) is a measure of the total amount of ozone over our heads; 1 DU is equivalent to an ozone thickness of 0.01 mm at 0°C and 1 atm pressure. What is the total mass of ozone present in the atmosphere if the average overhead amount is 200 DU? You may assume 0°C and 1 atm pressure.

11. As you know, the energy emitted by an object that has no internal energy (a so-called blackbody) as a function of wavelength is given by

$$E(\lambda) = \frac{2\pi hc^2 \lambda^{-5}}{e^{\frac{hc}{\lambda kT}} - 1}$$

where $E(\lambda)$ is the energy as a function of wavelength in W/m^3, λ is the wavelength (in m), h is Planck's constant (6.63×10^{-34} J s/molecule), c is the speed of light (3×10^8 m/s), k is the Boltzmann constant (1.38×10^{-23} J/deg), and T is the temperature of the object. Please use Excel to plot this function at four temperatures: 200, 700, 2000, and 6000 K between the wavelengths of 0.1 µm and 100 µm. Plot only energies above 10^{-4} W/m^2 µm (note the slight change in units). Put all four plots on the same set of axes. From these curves and the spreadsheet, read off the wavelengths at which the energy maximizes and report these wavelengths on your graph in the header information. Check these values with Wein's law. Please plot the above functions as lines with no symbols and be sure to label the axes using the proper units. *Hints*: Use about 350 rows for the wavelengths (in column A) incrementing them by a *factor* of 1.02 for each row. Use columns B – E for the four different temperatures. Be sure to plot everything on a log – log scale. If you use named variables, remember to substitute another name for the speed of light (Excel does not allow *c* as a variable name). Be careful with units—the above equation wants wavelengths in meters, but you may have them in microns. In the Excel Graph Wizard, be sure to use the "X-Y scatter plot" option – not the "line" option.

12. At a recent graduate student party, the hosts prepared a punch by adding 750 mL of vodka to sufficient fruit juice to bring the total volume up to 4 gal. The punch was consumed by the guests at a rate of one cup (there are 16 cups in a gallon) every 2 min. The hosts, however, replenished the punch by adding only fruit juice. In fact, with every cup of punch taken, the hosts added an equal volume of juice and mixed the bowl. At 11:30 P.M. noticing this

subterfuge, another graduate student (not one of the hosts) added 750 mL of vodka. Sketch a plot of the alcohol content (in percent) as a function of time between 8:30 P.M. (when the party started) and 2:30 A.M. (when the last guest left). Assume that vodka is 50% alcohol. Be sure to accurately show the half-life on your sketch.

13. The aquatic rate of the bacterial degradation of carbon tetrachloride depends on the concentration of trivalent iron in the solution:

$$CCl_4 + Fe^{3+} \rightarrow products + Fe^{2+}$$

A series of five experiments were carried out (each in its own reaction cell). In each of the five experiments, the concentration of iron was varied, and the concentration (in micromoles) of carbon tetrachloride in each reaction cell was measured as a function of time to give the following data.[14]

Time (h)	0 mM Fe^{3+}	5 mM Fe^{3+}	10 mM Fe^{3+}	20 mM Fe^{3+}	40 mM Fe^{3+}
0	19.5	19.5	19.5	19.5	19.5
1	19.5	18.3	20.0	17.7	16.5
6	19.5	18.7	17.0	14.6	12.1
16	19.5	16.2	13.1	11.8	7.2
30	19.5	13.7	12.5	7.8	3.5
42	19.5	11.3	8.4	4.5	0.8

---

[14] Thanks to Professor Flynn Picardal, School of Public and Environmental Affairs, Indiana University, for these data.

Clearly, the more iron, the faster the reaction goes. Your mission is to determine the second-order rate constant for this reaction. *Hint*: At each iron concentration, find the pseudo-first-order rate constant using the normal techniques. Then, remember that each of these pseudo-first-order rate constants is the product of the second-order rate constant and the iron concentration. You might fit a curve to the latter data as well. Be sure to get the units right.

14. Lake Philip is well mixed, and it has a volume of $10^8 \, m^3$. A single river flowing at $5 \times 10^5 \, m^3/day$ feeds it. Water exits Lake Philip through the Andrew River; evaporation is negligible. For several years, a local industry has been dumping 40 kg/day of dichloroastrene (DCTA) into Lake Philip. This chemical disappears from the lake by two processes: It flows out of the lake in the Andrew River, and it chemically degrades with a half-life of 48 days. (a) What is the concentration of DCTA in the lake? (b) If the flow of DCTA were suddenly reduced to 20 kg/day, how long would it take concentration to drop by one third?

15. What is the ozone concentration at 60 km altitude? You will need to know that the rate constants for the four Chapman equations vary with temperature as follows:

$$k_2 = 6.0 \times 10^{-34} \left(\frac{T}{300} \right)^{-2.3}$$

$$k_4 = 8.0 \times 10^{-12} \exp\left(\frac{-2060}{T} \right)$$

The values of k_1 and k_3 do not vary with temperature. At this altitude, the concentrations of NO and NO_2 are low enough to ignore.

16. An important reaction for the destruction of ozone is

$$Cl + O_3 \rightarrow ClO + O_2 \quad k(T) = 2.9 \times 10^{-11} e^{-260/T}$$

The rate constant for this reaction, in units of molecules^{-1} cm^3 s^{-1}, is given. (a) Near the Earth's equator, what is the rate of ozone destruction by this reaction at 30 km altitude, where the average concentration of Cl is about 4×10^3 cm^{-3}? (b) In the Antarctic ozone hole, the temperature is about $-80°C$, the concentration of ozone is about 2×10^{11} molecules/cm^3, and the concentration of atomic chlorine is about 4×10^5 molecules/cm^3. Under these conditions, what is the rate of ozone destruction for this reaction? What do you conclude? (c) What is the half-life of ozone in the Antarctic ozone hole?

17. During the deer hunting season in Michigan, the northernmost two thirds of the state host about 800,000 hunters. What is the average spacing between these hunters?[15]

18. While vacationing in Sweden one summer, Professor Hites, along with 10 of his closest friends, went on a hot-air balloon ride. The balloon had a volume of 310,000 ft^3, and together with the basket and burner, weighed 1000 kg. What was the air temperature inside the balloon?

---

[15] This question was designed in honor of Vice President Dick Cheney.

CHAPTER 4

CO$_2$ EQUILIBRIA

One motivation for studying CO_2 equilibria is to understand the effect that trace gases in the atmosphere have on the acidity (pH) of rain. Acid rain has been a problem of national and international scope with major economic consequences.

Our approach will be to set up several equations for various reactions and use them to find the pH of rain (or of a lake, etc.). First, we need to remember the definition of pH.

$$pH = -\log[H^+]$$

where log refers to the common (base 10) logarithm, and anything in square brackets refers to molar (moles per liter) concentration units. The use of the lowercase "p" here refers to the power of 10. It is true that

$$pANYTHING = -\log[ANYTHING]$$

For example, pK_a of acetic acid is 4.76, which means that K_a is $10^{-4.76} = 1.75 \times 10^{-5}$ (verify this for yourself).

Remember that the equilibrium constant for the reaction $A + B \leftrightarrow C + D$ is

$$K = \frac{[C][D]}{[A][B]}$$

Elements of Environmental Chemistry, by Ronald A. Hites
Copyright © 2007 John Wiley & Sons, Inc.

This means that when the reaction between A and B is at equilibrium (i.e., when the reaction has come to completion and when neither A or B is being lost nor C or D is being formed), the ratio of the concentrations of the products times each other divided by the concentrations of the reactants times each other is constant. If K is very small, then there are relatively low product concentrations compared to the reactant concentrations.[1] In fact, most interesting K values are usually small; hence, we use the pK notation. Also remember for pure water, the reaction

$$H_2O \leftrightarrow H^+ + OH^-$$

has an equilibrium constant of $10^{-14.00}$ at room temperature, or in our notation, p$K_w = 14.00$.

Finally, let us define the Henry's law constant, which is the ratio of the equilibrium concentration of a compound in solution to the equilibrium concentration of that compound in the gas phase over that solution. It is usually given as K_H. This constant can be given in one of two ways: with and without units. We will use the version with units here:

$$K_H = \frac{[X]}{P_X} = \frac{\text{Concentration of X in water (mol/L)}}{\text{Partial pressure of X in air over the water (atm)}}$$

In this case, the units of K_H are moles per liter atmosphere (mol L^{-1}atm^{-1}).

---

[1] Do not confuse the lower case k (a rate constant) with the upper case K (an equilibrium constant). Given that a reaction has had sufficient time to come to completion or to equilibrium, nothing is changing with time and the concepts of kinetics do not apply.

4.1   PURE RAIN

What is the pH of rain formed in and falling through the Earth's atmosphere if the atmosphere were free of anthropogenic pollutants (we might call this "pure rain")?

Strategy. The answer is not 7.00 as some might guess, but rather it is somewhat lower due to the presence of CO_2 in the atmosphere. The CO_2 dissolves into the rainwater, creates some carbonic acid (H_2CO_3), and lowers the pH of rain. Let us look at the reactions step by step. First, the CO_2 dissolves in the water. This is controlled by the K_H value, which is known experimentally:

$$CO_2(air) \leftrightarrow CO_2(water)$$

$$\frac{[CO_2]}{P_{CO_2}} = K_H = 10^{-1.47} M/atm$$

In some textbooks, the CO_2 dissolved in water is represented by H_2CO_3; this notation is chemically incorrect – H_2CO_3 is fully protonated carbonic acid. Another notation that you might encounter is $H_2CO_3{}^*$, which is the analytical sum of true H_2CO_3 and dissolved CO_2; at 25°C, dissolved CO_2 is 99.85% of this sum, so we will just use $[CO_2]$. Notice in this expression that the concentration of dissolved CO_2 is given in moles per liter (M) and the partial pressure is in atmospheres. We know that the atmospheric partial pressure of CO_2 is 380 ppm, which is 380×10^{-6} atm. Hence,

$$[CO_2] = 10^{-1.47} \times 380 \times 10^{-6} = 10^{-4.89} M$$

Next, we must consider the reactions of CO_2 with water:

$$CO_2 + H_2O \leftrightarrow HCO_3^- + H^+$$

HCO_3^- is called "bicarbonate." The above reaction has an equilibrium constant of

$$\frac{[HCO_3^-][H^+]}{[CO_2]} = K_{a1} = 10^{-6.35}$$

Rearranging this equation and substituting the CO_2 concentration from the Henry's law calculation above, we get

$$[HCO_3^-][H^+] = 10^{-6.35}[CO_2] = 10^{-6.35}10^{-4.89}$$

$$= 10^{-11.24}$$

$$[HCO_3^-] = \frac{10^{-11.24}}{[H^+]}$$

We are not done yet. There is another reaction in which bicarbonate dissociates to give carbonate and more acid:

$$HCO_3^- \leftrightarrow CO_3^{2-} + H^+$$

This reaction has the following equilibrium expression:

$$\frac{[CO_3^{2-}][H^+]}{[HCO_3^-]} = K_{a2} = 10^{-10.33}$$

Rearranging this expression and substituting the bicarbonate concentration from above, we get

$$[CO_3^{2-}][H^+] = 10^{-10.33}[HCO_3^-] = 10^{-10.33}\left(\frac{10^{-11.24}}{[H^+]}\right)$$

$$[CO_3^{2-}][H^+]^2 = 10^{-21.57}$$

$$[CO_3^{2-}] = \frac{10^{-21.57}}{[H^+]^2}$$

The dissociation of water is given by

$$[H^+][OH^-] = 10^{-14.00}$$

$$[OH^-] = \frac{10^{-14.00}}{[H^+]}$$

In the rainwater (or in any natural system), there must be the same concentration of negative charges as positive charges. This is called "charge balance." This is a very important concept. In this case, the charge balance is

$$[H^+] = [HCO_3^-] + 2[CO_3^{2-}] + [OH^-]$$

The "2" in front of the carbonate term is there because each mole of carbonate has 2 mol of charge. We can substitute concentrations from the above equations into the charge balance equation, taking care to eliminate all variables except $[H^+]$, and get

$$[H^+] = \frac{10^{-11.24}}{[H^+]} + \frac{2 \times 10^{-21.57}}{[H^+]^2} + \frac{10^{-14.00}}{[H^+]}$$

The OH^- term (the last one on the right) is about 600 $(= 10^{+2.76})$ times smaller than the $[HCO_3^-]$ term (the first on the right); hence, we will just drop the last term. Note that these two terms are of the same format, so it is simple to do this comparison. We also note that $2 = 10^{+0.30}$. Multiplying through by $[H^+]^2$ gives us

$$[H^+]^3 = 10^{-11.24}[H^+] + 10^{-21.27}$$

If we guess that the pH is about 6, we can test the remaining terms to see if any are too small to keep. In this case, we get

$$10^{-18} = 10^{-17.2} + 10^{-21.3}$$

This indicates that the last term on the right is about 2000 times smaller than the others and can be neglected. The final equation is

$$[H^+]^2 = 10^{-11.24}$$
$$[H^+] = 10^{-11.24/2} = 10^{-5.62}$$
$$pH = -\log[H^+] = 5.62$$

Hence, the pH of pure rain is 5.62, which agrees well enough with our guess.[2]

Using a pH of 5.62, calculate the concentrations of each charged species using the equilibrium expressions and then check that the rainwater is electrically neutral. Are we justified in omitting the two rightmost terms?

---

[2] For future reference, at an atmospheric CO_2 concentration of 380 ppm, $[CO_2] = 10^{-4.89}$; $[HCO_3^-] = 10^{-11.24}/[H^+]$; and $[CO_3^{2-}] = 10^{-21.57}/[H^+]^2$.

Strategy. The four terms are

$$[H^+] = 10^{-5.62} = 2.4 \times 10^{-6} \, M$$

$$[HCO_3^-] = \frac{10^{-11.24}}{10^{-5.62}} = 10^{-5.62} = 2.4 \times 10^{-6} \, M$$

$$2[CO_3^{2-}] = \frac{10^{+0.30} 10^{-21.57}}{10^{-5.62 \times 2}} = 10^{-10.03} = 9.3 \times 10^{-11} \, M$$

$$[OH^-] = \frac{10^{-14.00}}{10^{-5.62}} = 10^{-8.38} = 4.2 \times 10^{-9} \, M$$

The only two species that contribute significantly to the charge are H^+ and HCO_3^-. Thus, charge balance is achieved when $[H^+] = [HCO_3^-]$ at pH = 5.62.

4.2 POLLUTED RAIN

What would the pH of rain be if the atmosphere also had 0.5 ppb of SO_2 in it?

Strategy. In this case, we need another set of reactions for the solution of SO_2 from the gas phase into the water (rain) and for the reactions of SO_2 with water. These are just like the CO_2 reactions except that they have different equilibrium constant (K) values.

$$SO_2 \, (air) \leftrightarrow SO_2 \, (water)$$

$$\frac{[SO_2]}{P_{SO_2}} = K_H = 10^{+0.096} \, M/atm$$

This is the measured Henry's law constant for SO_2; hence, the pK_H for SO_2 is -0.096.

We are given that the atmospheric partial pressure of SO_2 is 5×10^{-10} atm. Hence,

$$[SO_2] = 10^{+0.096} \times 5 \times 10^{-10} = 10^{-9.21} \text{ M}$$

We must consider the reactions of SO_2 with water:

$$SO_2 + H_2O \leftrightarrow HSO_3^- + H^+$$

which has an equilibrium constant of

$$\frac{[HSO_3^-][H^+]}{[SO_2]} = K_{a1} = 10^{-1.77}$$

HSO_3^- is called "bisulfite." Rearranging this equation and substituting the SO_2 concentration from the Henry's law calculation above, we get

$$[HSO_3^-][H^+] = 10^{-1.77}[SO_2] = 10^{-1.77}10^{-9.21}$$

$$= 10^{-10.98} \text{M}^2$$

$$[HSO_3^-] = \frac{10^{-10.98}}{[H^+]}$$

We are not done yet. There is another reaction in which bisulfite dissociates to give more acid:

$$HSO_3^- \leftrightarrow SO_3^{2-} + H^+$$

which has the following equilibrium expression:

$$\frac{[SO_3^{2-}][H^+]}{[HSO_3^-]} = K_{a2} = 10^{-7.21}$$

SO_3^{2-} is called sulfite. Rearranging this expression and substituting the bisulfite concentration from above, we get

$$[SO_3^{2-}][H^+] = 10^{-7.21}[HSO_3^-] = 10^{-7.21}\left(\frac{10^{-10.98}}{[H^+]}\right)$$

$$[SO_3^{2-}][H^+]^2 = 10^{-18.19}$$

$$[SO_3^{2-}] = \frac{10^{-18.19}}{[H^+]^2}$$

The revised charge balance is

$$[H^+] = [HCO_3^-] + 2[CO_3^{2-}] + [HSO_3^-] + 2[SO_3^{2-}] + [OH^-]$$

The "2" in front of the sulfite term is there because each mole of sulfite has 2 mol of charge. We can substitute concentrations from the above equations into the charge balance equation, taking care to eliminate all variables except $[H^+]$, and get

$$[H^+] = \frac{10^{-11.24}}{[H^+]} + \frac{10^{+0.30}10^{-21.57}}{[H^+]^2} + \frac{10^{-10.98}}{[H^+]}$$
$$+ \frac{10^{+0.30}10^{-18.19}}{[H^+]^2} + \frac{10^{-14.00}}{[H^+]}$$

If we guess that the pH is about 5, we can test the remaining terms to see if any are too small to keep. We get

$$10^{-5} = 10^{-6.2} + 10^{-11.3} + 10^{-6.0} + 10^{-7.9} + 10^{-9}$$

This suggests that we should keep only the first and third terms on the right. This gives

$$[H^+]^2 = 10^{-11.24} + 10^{-10.98} = 10^{-10.79}$$

$$[H^+] = 10^{-10.79/2} = 10^{-5.39}$$

$$pH = -\log[H^+] = 5.39$$

Hence, the pH of rain falling through air with 0.5 ppb of SO_2 in it is much more acidic than without the SO_2. In this case, the $[H^+]$ is about 70% higher with SO_2 than without it.[3]

What would the pH of rain be if the atmospheric concentration of SO_2 were 10 times higher than the background value?

Strategy. We can use the same reactions and Henry's law constant and equilibrium constants, but we need to change the partial pressure of SO_2 to

$$10 \times 5 \times 10^{-10} = 5 \times 10^{-9} \, atm.$$

Hence,

$$[SO_2] = 10^{+0.096} \times 5 \times 10^{-9} = 10^{-8.21} \, M$$

Substituting this concentration into the first equilibrium expression, we get

---

[3] For future reference, at an atmospheric SO_2 concentration of 0.5 ppb, $[SO_2] = 10^{-9.21}$, $[HSO_3^-] = 10^{-10.98}/[H^+]$, and $[SO_3^{2-}] = 10^{-18.19}/[H^+]^2$.

$$[HSO_3^-][H^+] = 10^{-1.77}[SO_2] = 10^{-1.77}10^{-8.21}$$
$$= 10^{-9.98} \, M^2$$
$$[HSO_3^-] = \frac{10^{-9.98}}{[H^+]}$$

Given that this rain is even more acidic, we only need to retain the same three terms of the charge balance equation that we had before:

$$[H^+] = [HCO_3^-] + [HSO_3^-]$$

Substituting concentrations from the above equations into the charge balance equation and taking care to eliminate all variables except $[H^+]$, we get

$$[H^+] = \frac{10^{-11.24}}{[H^+]} + \frac{10^{-9.98}}{[H^+]}$$

This gives

$$[H]^2 = 10^{-11.24} + 10^{-9.98} = 10^{-9.96}$$

Notice that with this much SO_2 in the atmosphere, the acid contribution from CO_2 is very small; in other words, the bicarbonate term ($10^{-11.24}$) is about 20 times smaller than the bisulfite term ($10^{-9.98}$). We can get the pH from the above equation.

$$[H^+] = 10^{-9.96/2} = 10^{-4.98}$$
$$pH = -\log[H^+] = 4.98$$

This is a lot of acid and not that much SO_2. Thus, it is easy to see that control of acid rain relies on control of sulfur emissions (mostly) from burning high-sulfur coal.[4]

By the way, these dissolution and acid forming reactions of SO_2 in rainwater are only the first step in forming acid rain. The last step is the oxidation of these sulfur species to form sulfate ions (SO_4^{2-}). Thus, the form of the acid in the rain water is sulfuric acid (H_2SO_4), which is a stronger acid than sulfurous acid (H_2SO_3), and which tends to reduce the pH of the rain even more.

Let us go back to an SO_2 concentration of 0.5 ppb and add some ammonia to the atmosphere at a concentration of 0.02 ppb. Now what would the pH of rain be?

Strategy. The equilibrium reaction for ammonia with water is

$$NH_3 + H_2O \leftrightarrow NH_4^+ + OH^-$$

The equilibrium constant is known to have a $pK_b = 4.74$,[5] hence

$$[NH_4^+][OH^-] = 10^{-4.74}[NH_3]$$

---

[4] For future reference, at an atmospheric SO_2 concentration of 5 ppb, $[SO_2] = 10^{-8.21}$ and $[HSO_3^-] = 10^{-9.98}/[H^+]$.

[5] Unlike the equilibria constants for CO_2 and SO_2, which are abbreviated as K_a, where "a" stands for acid, this equilibrium constant is for a base and is abbreviated as K_b, where "b" stands for base, but it is basically the same concept.

Ammonia is very water soluble, and its Henry's law constant is known to have a $pK_H = -1.76$; hence

$$[NH_3] = 10^{+1.76}P_{NH_3}$$

In this problem, the partial pressure of ammonia is $10^{-10.70}$ atm, given a concentration of 0.02 ppb.

$$[NH_3] = 10^{+1.76}10^{-10.70} = 10^{-8.94}$$

Omitting the terms we know we will not need, the charge balance equation is

$$[H^+] + [NH_4^+] = [HCO_3^-] + [HSO_3^-]$$

We already know the two terms on the right

$$[HCO_3^-] = \frac{10^{-11.24}}{[H^+]}$$

$$[HSO_3^-] = \frac{10^{-10.98}}{[H^+]}$$

but $[NH_4^+]$ is a new term on the left. Putting the equilibrium expression and the Henry's law constant for ammonia together, we have

$$[NH_4^+] = \frac{10^{-4.74}10^{-8.94}}{[OH^-]} = \frac{10^{-13.68}}{[OH^-]}$$

We can always use the equilibrium expression for water to get

$$[OH^-] = \frac{10^{-14.00}}{[H^+]}$$

Therefore,

$$[NH_4^+] = 10^{-13.68}10^{+14.00}[H^+]$$

$$= 10^{+0.32}[H^+] = 2.09[H^+]$$

Now we can put all of this stuff into the charge balance equation, and we get

$$[H^+] + 2.09[H^+] = \frac{10^{-11.24} + 10^{-10.98}}{[H^+]}$$

This is an equation with one unknown value, which we can solve.

$$[H^+] = \left(\frac{10^{-11.24} + 10^{-10.98}}{3.09}\right)^{1/2}$$

$$= \left(\frac{10^{-10.79}}{10^{+0.49}}\right)^{1/2} = 10^{-5.64}$$

$$pH = -\log[H^+] = 5.64$$

This is actually quite remarkable—by adding just a little bit of a very water-soluble base (NH_3 in this case), the pH of the rain goes almost back to that of pure rain.

How much NH_3 would it take to offset the acid caused by an SO_2 atmospheric concentration of 5 ppb?

Strategy. We are aiming for a pH of 5.62. Given that we know all of the equilibria expressions and the charge balance, we just need to write everything leaving the partial pressure of NH_3 as the unknown. We know the bicarbonate term is $10^{-11.24}/[H^+]$ because the atmospheric CO_2 concentration has not changed, and we know the bisulfite term is $10^{-9.98}/[H^+]$ because we just figured that out for this SO_2 concentration (see the second problem in this section). We also know that the ammonium term is

$$[NH_4^+] = \frac{K_H P_{NH_3} K_b [H^+]}{K_w}$$

Thus, the charge balance equation is

$$[H^+] + [NH_4^+] = [HCO_3^-] + [HSO_3^-]$$

$$[H^+] + \frac{K_H P_{NH_3} K_b [H^+]}{K_W}$$

$$= \frac{10^{-11.24}}{[H^+]} + \frac{10^{-9.98}}{[H^+]} = \frac{10^{-9.96}}{[H^+]}$$

$$[H^+]^2 (1 + 10^{-4.74} 10^{+1.76} 10^{+14.00} P_{NH_3}) = 10^{-9.96}$$

At pH $= 5.62$

$$1 + 10^{+11.02} P_{NH_3} = 10^{-9.96 + 2 \times 5.62} = 10^{+1.28}$$

$$P_{NH_3} = (10^{+1.28} - 1) \times 10^{-11.02}$$

$$= 10^{1.26} 10^{-11.02}$$

$$= 10^{-9.76} = 1.72 \times 10^{-10} \text{atm}$$

$$= 0.17 \text{ppb}$$

Although ammonia concentrations vary a lot from place to place, 0.17 ppb is not an unrealistic concentration.

4.3 SURFACE WATER

What is the pH of the water in an Indiana limestone quarry?

Strategy. Remember that limestone is $CaCO_3$ and when it dissolves in water it dissociates:

$$CaCO_3(s) \leftrightarrow Ca^{2+} + CO_3^{2-}$$

The degree of dissociation is given by the solubility product constant

$$K_{sp} = [Ca^{2+}][CO_3^{2-}] = 10^{-8.42}$$

Note that the solubility product constant expression is true *only* when there is undissolved, solid material still present in the system. In other words, we are talking about (in this case) solid calcium carbonate in equilibrium with a saturated solution of calcium carbonate.

The charge balance equation for the water in the quarry is a little different from that of the rain because it now includes calcium ions:

$$[H^+] + 2[Ca^{2+}] = [HCO_3^-] + 2[CO_3^{2-}] + [OH^-]$$

Note the factor of 2 on the calcium concentration. Why is it there?

From the carbonate equilibrium expression derived previously, we know that

$$[CO_3^{2-}] = \frac{10^{-21.57}}{[H^+]^2}$$

Hence, from the K_{sp} equation

$$[Ca^{2+}] = 10^{-8.42}10^{+21.57}[H^+]^2 = 10^{+13.15}[H^+]^2$$

We can now substitute this and the other CO_2 equations into the charge balance equation and get

$$H^+] + 2 \times 10^{+13.15}[H^+]^2 = \frac{10^{-11.24}}{[H^+]} + \frac{2 \times 10^{-21.57}}{[H^+]^2} + \frac{10^{-14.00}}{[H^+]}$$

Let us guess that the pH of the lake is 7 and see how big the terms are.

$$10^{-7} + 10^{-0.6} = 10^{-4.2} + 10^{-7.3} + 10^{-7.0}$$

This suggests that we keep only the second term on the left and the first term on the right

$$10^{+13.45}[H^+]^2 = \frac{10^{-11.24}}{[H^+]}$$

Checking the full charge balance equation, we note that we have kept only the calcium and the bicarbonate terms. This suggests that the water in the quarry is mostly a solution of calcium bicarbonate.

The solution to this simplified equation is

$$[H^+]^3 = 10^{-11.24}10^{-13.45} = 10^{-24.69}$$
$$[H^+] = 10^{-24.69/3} = 10^{-8.23}$$

Thus, the pH is 8.23, which agrees well with observations. How did the pH of the lake get to be so much higher than the pH of the rain falling into it?

It is important to remember that this last calculation assumed that solid calcium carbonate was present in the system; for example, the lake bed was limestone. It should be clear that the partial pressure of CO_2 over a lake cannot usually exceed about 380 ppm, the global average atmospheric concentration, and thus, the calcium concentration cannot exceed a certain level if that partial pressure is to be maintained. In the above calculation, remember that the exponent -11.24 was based on a CO_2 partial pressure of 380 ppm. At this pressure and at a pH of 8.23 (which is what we just calculated), the calcium concentration is given by

$$[Ca^{2+}] = 10^{+13.15}[H^+]^2 = 10^{+13.15}10^{-2\times8.23}$$
$$= 10^{-3.31} = 4.9 \times 10^{-4} \, M$$

In other words, at 380 ppm of CO_2, the maximum dissolved calcium concentration is about 500 µM. Of course, it can be less if all of the solid calcium carbonate is dissolved, and it can be more if the pressure of CO_2 is higher (as it might be in a groundwater sample or in a closed can of a carbonated beverage). Let us ask a more general question.

What is the solubility of calcium in water as a function of the partial pressure of CO_2 in equilibrium with that water?

Strategy. By definition, we know that

$$[Ca^{2+}][CO_3^{2-}] = K_{sp} = 10^{-8.42}$$

and

$$[CO_2] = K_H P_{CO_2}$$

and

$$[HCO_3^-][H^+] = K_{a1}[CO_2] = K_{a1}K_H P_{CO_2}$$

Using a simplified charge balance, we have

$$2[Ca^{2+}] = [HCO_3^-]$$

Substituting this into the K_{a1} expression, we have

$$2[Ca^{2+}][H^+] = K_{a1}K_H P_{CO_2}$$

which rearranges to

$$[H^+] = \frac{K_{a1}K_H P_{CO_2}}{2[Ca^{2+}]}$$

Substituting this into the K_{a2} expression

$$[CO_3^{2-}][H^+] = K_{a2}[HCO_3^-]$$

we have

$$[CO_3^{2-}] = \frac{K_{a2}[HCO_3^-] \times 2[Ca^{2+}]}{K_{a1}K_H P_{CO_2}}$$

From the charge balance equation, we have $[HCO_3^-] = 2[Ca^{2+}]$. Substituting this in the above expression, we have

$$[CO_3^{2-}] = \frac{4[Ca^{2+}]^2 K_{a2}}{K_{a1}K_H P_{CO_2}}$$

Substituting this into the K_{sp} expression, we have

$$[Ca^{2+}]\left(\frac{4[Ca^{2+}]^2 K_{a2}}{K_{a1}K_H P_{CO_2}}\right) = K_{sp}$$

which rearranges to

$$[Ca^{2+}]^3 = \left(\frac{K_{sp}K_{a1}K_H}{4K_{a2}}\right)P_{CO_2}$$

or

$$[Ca^{2+}] = \left(\frac{K_{sp}K_{a1}K_H}{4K_{a2}}\right)^{1/3} P_{CO_2}^{1/3}$$

The values of the various equilibrium constants are known; therefore, this equation becomes

$$[Ca^{2+}]_{max} = 0.00675 P_{CO_2}^{1/3}$$

You can plot this for yourself using Excel, but a few benchmarks are handy. At 380 ppm of CO_2, the maximum dissolved calcium concentration is 490 μM or 20 mg/L; at 0.1 atm, it is 3100 μM or 125 mg/L; and at 0.5 atm, it is 5400 μM or 215 mg/L.

4.4 PROBLEM SET

1. For a lake in northern Florida, plot the logarithm of the following versus pH in the range of 0-10 pH units: P_{CO_2}, $[CO_2]$, $[HCO_3^-]$, $[H^+]$, $[OH^-]$, and $[CO_3^{2-}]$. Using the equations for the lines from which you constructed this graph, answer the fol-

lowing: At what pH would the carbonate concentration be the same as the hydroxide concentration? At what pH would the dissolved CO_2 concentration equal the carbonate concentration? At what pH would the carbonate concentration just start to exceed the bicarbonate concentration?

2. What would be the pH of a soda water (e.g., a Coke) made by saturating pure water with pure CO_2 at 1 atm pressure?

3. Estimate the calcium concentration in a groundwater sample, which has a pH of 5.50. Assume that the groundwater at this location is saturated with CO_2 at a partial pressure of 0.1 atm.

4. What is the solubility of oxygen in lake water at 28°C? Assume that the pK_H of oxygen at this temperature is twice that of CO_2. Give your answer in mg/L.

5. The calcium concentration of Lake Mary (a lake in New Hampshire) is 4×10^{-4} M. Estimate the pH of this lake. Assume that calcium inputs to this lake are exclusively from calcium carbonate weathering.

6. The drinking water for Bloomington, Indiana, comes from Lake Monroe, which has a calcium concentration of 17 ppm. What is the pH of this lake? It may (or may not) help to remember that the atomic weight of calcium is 40.1 g/mol.

7. A water sample has a pH of 8.44 and a total calcium concentration of 1.55 ppm. For this question, assume that the only ions present in the water are Ca^{2+}, HCO_3^-, and CO_3^{2-}. What are the concentrations of CO_3^{2-} and HCO_3^- in moles per liter?

8. Before the industrial revolution, the concentration of CO$_2$ in the Earth's atmosphere was about 275 ppm. Considering the effect of dissolved CO$_2$ only, calculate the effect that the increase in CO$_2$ has had on the pH of precipitation.

9. A sample of rainwater is observed to have a pH of 7.4. If only atmospheric CO$_2$ at 380 ppm and limestone dust are present in the atmosphere to alter the pH from a neutral value, and if each raindrop has a volume of 0.02 cm^3, what mass of calcium is present in each raindrop?

10. The concentration of pentachloroamylene (PCA) in Lake Henry is 3.2 ng/L, which has an average depth of 25 m. PCA is removed from this lake only by deposition to the sediment, and the rate constant for this process is $2.1 \times 10^{-4}\,h^{-1}$. The only source of PCA to this lake is rain. What is the concentration of PCA in the rainwater (in ng/L)? Assume that the precipitation rate is 80 cm/year.

11. While in Italy one summer, Professor Hites ordered a bottle of local mineral water with dinner. It had a pH of about 8, and the following composition was printed on the label: Na$^+$, 47 mg/L; K$^+$, 46 mg/L; Mg^{2+}, 19 mg/L; HCO$_3^-$, 1397 mg/L; Cl$^-$, 23 mg/L; and NO$_3^-$, 5.5 mg/L. The Ca^{2+} concentration was illegible. What was the Ca^{2+} concentration in mg/L? You may (or may not) need the following atomic weights: Ca, 40.1; Na, 23.0; K, 39.1; Mg, 24.3; and Cl, 35.5.

12. Imagine that a polluter starts dumping sodium chloride into Lake Charles at a rate of 1600 kg/day, that the background concentration of NaCl in the

lake is 11 ppm, and that the residence time of NaCl in the lake is 3.5 years. After 5 years, the Environmental Protection Agency catches on and turns off this source of NaCl. What would be the maximum concentration of NaCl in the lake? Please give your answer in ppm. You may (or may not) need the following facts: Lake Charles has a volume of $1.8 \times 10^7 \, m^3$. The density of NaCl is twice that of water.

13. Ethane (C_2H_6) makes up about 6% of natural gas. Ethane is only emitted into the atmosphere whenever natural gas escapes unburned at wells and from leaking pipelines. The average concentration of ethane in the troposphere in the Northern Hemisphere is about 1.0 ppb, and in the Southern Hemisphere, it is about 0.5 ppb. Ethane can exit from the troposphere in three ways: passage to the stratosphere, chemical reactions in the troposphere, and wet deposition to the Earth's surface. Ethane can also leave one hemisphere by flowing into the other. Assume that all of these exit processes are of first order and that all of the sources are in the Northern Hemisphere. Our best guess is that 3% as much natural gas escapes to the atmosphere as is burned and that about $1.5 \times 10^{12} \, m^3$ of natural gas is burned annually. Please estimate the net rate of ethane flow across the equator.[6]

---

[6] This problem, but not the solution, was from *Consider a Spherical Cow* by J. Harte; used with permission.

14. A student was sent to a house (volume $= 21{,}000$ ft^3) to measure the ventilation rate of the house. She quickly added enough SF$_6$ (a nontoxic, inert gas) to the indoor air to bring its concentration up to 100 ppb. She then measured the concentration of this compound every 6 min for about 5 h. The results are shown below. How many air exchanges per hour does this house have? What happened at point A?

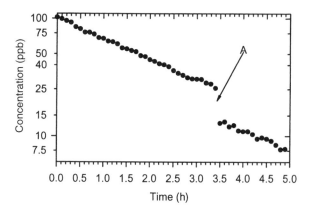

15. Certain types of cigarettes give off smoke laden with the chemical THC, which can reach concentrations of $200 \, \mu g/m^3$ in some Amsterdam coffeehouses. Although only 10% of the THC breathed into the lungs actually enters the bloodstream, patrons intake a substantial amount of the chemical. Assume that the average breathing rate is 20 L/min and the residence time of the THC in the body is 6 h. (a) What is the steady-

state concentration (in ppb) of the THC in the body of an American tourist who never leaves the coffeehouse? (b) What concentration has the chemical reached after the tourist has been in the coffeehouse for just 3 h? Assume that this average tourist weighs 70 kg.

16. If the entire population of the planet were to die and be set afloat in Lake Michigan (I know this is morbid, but it is just pretend), how much higher would the water level rise? The area of Lake Michigan is 22,000 square miles.

17. A student in an atmospheric chemistry laboratory had measured the rate constant for the reaction of isoprene (C_5H_8) with OH in a small chamber with a volume of 200 cm^3. Her result was 9.4×10^{-11} cm^3/s. Later the question came up: What was the steady-state concentration of OH in this chamber? The student went back to her original data records and noticed that the concentration of isoprene had decreased by 25% in 3 min. She had injected 2 µL of a solution of isoprene in CCl$_4$ into the chamber, and the concentration of this solution was 6 µg/µL. From these data, she found an answer; can you?

18. Lake Titicaca is situated at an altitude of 3810 m in the Bolivian Andes. Calculate the solubility of oxygen in this lake at a temperature of 5°C. The Henry's law constant at this temperature is 1.9×10^{-8} mol L^{-1} Pa^{-1}. "Pa" here refers to pascal, which is the official SI unit of pressure.

19. At sea level and at 30°C, the solubility of oxygen in water is 7.5 mg/L. Consider a water body at that

temperature containing 7.0 mg/L of oxygen. By photosynthesis, 1.5 mg/L of CO$_2$ is converted to organic biomass, which has a composition of C$_6$H$_{12}$O$_6$, during a single hot day. Is the amount of oxygen produced at the same time sufficient to exceed its aqueous solubility?

CHAPTER 5

FATES OF ORGANIC COMPOUNDS

What happens to an organic compound when it is dumped into the environment? Clearly, the answer depends on the physical and chemical properties of the compound. For example, a big spill of methane will not cause a water pollution problem, but a major release of DDT could cause a big problem for biota. This chapter will provide a few tools for looking quantitatively at these issues. Of necessity, we will be brief, but for a more complete coverage, the reader is referred to the massive book by Schwarzenbach et al.[1]

We will address the distribution of organic compounds in the environment by looking at equilibrium partitioning of organic compounds between environmental "phases," which include air, water, soil, and biota. Taking these phases pairwise, we can define the various physical and chemical properties that control the partition coefficients between these phases:

---

[1] R.P. Schwarzenbach, P.M. Gschwend and D.M. Imboden, *Environmental Organic Chemistry*, 2nd edition, Wiley–Interscience, Hoboken, NJ, 1313 pp.

Elements of Environmental Chemistry, by Ronald A. Hites
Copyright © 2007 John Wiley & Sons, Inc.

air–water	vapor pressure and water solubility
water–soil	adsorption and water solubility
soil–air	adsorption and vapor pressure
all phases	solubility in fats
with biota	(called "lipophilicity")

If these various physical and chemical properties are known, then we can predict in which phase (and to what extent) an organic compound would end up. Let us look at these properties one by one.

5.1   VAPOR PRESSURE

Vapor pressure is essentially the solubility of a compound in air. Permanent gases, such as methane, have high vapor pressures; in fact, they have a vapor pressure of 1 atmosphere (atm) or 760 Torr. Some pesticides have medium vapor pressures; for example, hexachlorobenzene has a vapor pressure of about 10^{-7} atm. Some compounds, such as decachlorobiphenyl, have vapor pressures that are so low that they are essentially nonvolatile (10^{-10} atm). For our purposes, the interesting range is 10^{-4} to 10^{-8} atm.

Vapor pressures are a strong function of temperature. The relationship is given by the Clausius–Clapeyron equation:

$$\ln(P) = -\frac{\Delta H_{vap}}{R}\left(\frac{1}{T}\right) + \text{const}$$

where ΔH_{vap} is the compound's heat of vaporization, R is the gas constant (8.3 J deg^{-1} mol^{-1}), T is the temperature of the system, and "const" is a constant depending on the compound. To use this equation, one must know at least one vapor pressure at one temperature and the compound's ΔH_{vap} value. The latter is usually in the range of 50–90 kJ/mol. A common use of this equation is for plotting ambient atmospheric concentrations of a compound as a function of the atmospheric temperature when the sample was taken. In this case, a plot of $\ln P$ versus $1/T$ should be linear with a slope of $-\Delta H/R$.

If we know a compound's boiling point (which is the temperature at which its vapor pressure is 1 atm), we can predict its vapor pressure at a given temperature (T) from

$$\ln P_L = -(4.4 + \ln T_B)\left[1.8\left(\frac{T_B}{T} - 1\right) - 0.8\ln\left(\frac{T_B}{T}\right)\right]$$

where P_L is the vapor pressure (in atm) of the subcooled liquid (a liquid even at a temperature below the compound's melting point) and T_B is the boiling point in K. The vapor pressure of a solid (P_s) is given by

$$\ln\left(\frac{P_S}{P_L}\right) = -6.8 \quad \text{or} \quad P_S = 0.0011\,P_L$$

5.2   WATER SOLUBILITY

This refers to the saturated solubility of the organic compound in water. We will give it the symbol of C_w^{sat}, which is in mol/L. For typical organic pollutants, this value is usually very low. For example, a relatively high value might be for benzene (0.1 mol/L), and a relatively low value might be for decachlorobiphenyl (10^{-10} mol/L). C_w^{sat} changes with temperature but not as much as vapor pressure. Unfortunately, there are no good prediction methods.

5.3   HENRY'S LAW CONSTANT

As we have seen in the previous section, the Henry's law constant is the ratio of a compound's water solubility to its partial pressure above that water sample. In this section, we will use the symbol H for the Henry's law constant (or sometimes HLC).

$$H = \frac{P_L}{C_w^{sat}}$$

Note that the units of H are atm L/mol and that

$$H = \frac{1}{K_H}$$

For CO_2, H would be $10^{+1.47}$. See the previous section for the use of K_H. A unitless HLC is sometimes used as well:

$$H' = \frac{H}{RT} = \frac{C_{air}}{C_{water}}$$

Like vapor pressure, the HLC is a strong function of temperature. Unfortunately, there are no good predictive methods for the HLC.

What is the unitless Henry's law constant for CO_2 at 25°C?

Strategy. From the previous chapter, we know that $K_H = 10^{-1.47}$ mol/L atm at "room temperature" (300 K); thus,

$$H' = \frac{H}{RT} = \frac{1}{K_H RT} = \left(\frac{L\,atm}{10^{-1.47}\,mol}\right)\left(\frac{deg\,mol}{0.082\,L\,atm}\right)$$
$$\times\left(\frac{1}{298\,deg}\right) = 1.21$$

5.4   PARTITION COEFFICIENTS

Partition coefficients are the ratio of the concentration of an organic compound in two phases that are in equilibrium with each other. Imagine a separatory funnel with an organic solvent layer on the top and water at the bottom. Compound X is in one or the other of the layers. After you have carefully shaken the funnel and waited for the phases to separate, you then measure the concentration of compound X in both phases. The partition coefficient is

$$K = \frac{C_{organic}}{C_{water}}$$

Note that the denominator concentration will be something less than C_w^{sat}. A high value of K suggests that the compound is not very water soluble but is more soluble in organic solvents. In this case, the compound is said to be lipophilic (or fat soluble).

5.5 LIPOPHILICITY

To simulate lipids (fats) in biota, pharmacologists long ago selected a model compound: *n*-octanol. Thus, the partition coefficient that best describes lipophilicity is the octanol–water partition coefficient, usually given the symbol K_{ow}. The interesting values of K_{ow} are usually in the 10^2–10^7 range; hence, it is convenient to use the common logarithm of K_{ow}.

K_{ow} is related to water solubility; high water solubility implies a low K_{ow}. In fact, there is an empirical relationship between these two parameters:

$$\log K_{ow} = -0.86 \log C_w^{sat} + 0.32$$

Remember that C_w^{sat} is in mol/L.

If one knows the $\log K_{ow}$ value of a given compound, it is possible to calculate the $\log K_{ow}$ value of a related compound by adding or subtracting "pi-values." The pi-values for common substituents are as follows.

Pi-values			
NH_2	−1.23	F	0.14
OH	−0.67	$N(CH_3)_2$	0.18
CN	−0.57	CH_3	0.56
NO_2	−0.28	Cl	0.71
COOH	−0.28	Br	0.86
OCH_3	−0.02	C_2H_5	0.98
H	0.00	$CH(CH_3)_2$	1.35

Let us take an example: Assume that you know the $\log K_{ow}$ value for a trichlorobiphenyl is 6.19, and you want to estimate the log K_{ow} for a tetrachlorobiphenyl. Simply add the pi-value for chlorine (0.71) to the base log K_{ow} value and get $6.19 + 0.71 = 6.90$.

The structures of DDT and methoxychlor are given below (left and right, respectively). Given that the log K_{ow} value for DDT is 5.87, what is the log K_{ow} value of methoxychlor? By what factor is it more or less lipophilic than DDT?

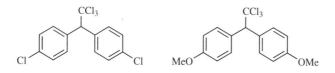

Strategy. The only structural difference between these two molecules is the replacement of two chlorine atoms on DDT with two methoxy (also written as OCH_3) moieties on methoxychlor. Thus, we can use the pi-values of these two groups to first remove the two chlorines and then add the two methoxys.

Methoxychlor's   log K_{ow} value $= 5.87 - 2 \times 0.71 + 2 \times (-0.02) = 4.41$. The K_{ow} values are the antilogarithms of these two values: for DDT, $K_{ow} = 10^{5.87} = 7.41 \times 10^5$, and for methoxychlor, $K_{ow} = 10^{4.41} = 2.57 \times 10^4$. Because it has a higher K_{ow} value, DDT is the more lipophilic of the two compounds, and it is more lipophilic by a factor of

$$\frac{7.41 \times 10^5}{2.57 \times 10^4} = 29$$

We could actually solve this sort of problem with just the pi-values and ignore the rest of the molecules. Notice that the *difference* in the two $\log K_{ow}$ values is $2 \times 0.71 - 2 \times (-0.02) = 1.46$. Because this is a difference in logarithms, the antilogarithm (or exponentiation) of this difference is the factor by which the two values differ.[2] Hence,

$$\log K_{ow}(\text{DDT}) - \log K_{ow}\ (\text{methoxychlor})$$
$$= 2 \times 0.71 + 2 \times 0.02 = 1.46$$

$$\frac{K_{ow}(\text{DDT})}{K_{ow}(\text{methoxychlor})} = 10^{1.46} = 29$$

5.6 FISH PARTITION COEFFICIENTS

The simplest case of an animal in equilibrium with its surroundings is a fish. In this case, we can define a partition coefficient for the concentration of some organic compound in the fish relative to the concentration of that compound in the water in which the fish lives.

$$K_B = \frac{C_{\text{fish}}}{C_{\text{water}}}$$

where C_{fish} is the concentration of some pollutant in the whole fish, usually in units of µg/g wet weight of fish, and C_{water} is the concentration of the same pollutant in the surrounding water, usually in units of µg/cm^3 of water. Given the density of water, K_B has no units.

---

[2] Remember the property of logarithms: $\log(A) - \log(B) = \log\frac{A}{B}$

Because octanol was selected to simulate animal lipids, K_B is related to K_{ow}; the empirical relationship is

$$\log K_B = \log K_{ow} - 1.32$$

or exponentiating both sides

$$K_B = 0.048\, K_{ow}$$

5.7 ADSORPTION

Adsorption refers to the collection of organic compounds on the surfaces of particles, such as soil or suspended sediment. Most of these particles are covered with a layer of organic material; thus, the adsorption results from the attraction of two organic materials for one another. Adsorption is measured by a partition coefficient, which is the ratio of the concentration of the compound on the solid to its concentration in the water surrounding the solid:

$$K_d = \frac{C_{\text{solid}}}{C_{\text{water}}}$$

The concentration on the solid has units of mol/kg, and the concentration in the water is mol/L; hence, K_d has units of L/kg. Assuming a solid density of 1 kg/L, these units are often ignored.

K_d will often depend on how much of the total mass of the particle is organic material; thus, K_d can be corrected by the fraction of organic material (f_{om}) in the particles to give

$$K_{\text{om}} = \frac{K_d}{f_{\text{om}}}$$

Note that f_{om} is a fraction and is always less than 1 and often less than 0.1. Because the partitioning is from the water to the organic material on the particle, it should come as no surprise that K_{om} is empirically related to K_{ow} and to the water solubility:

$$\log K_{om} = +0.82 \log K_{ow} + 0.14$$
$$\log K_{om} = -0.75 \log C_w^{sat} + 0.44$$

A covered soup blow contains 1 L of a very dilute, cold soup (at 25°C), 1 L of air, and a floating blob of fat with a volume of 1 mL. The system also contains 1 mg of naphthalene. The $\log K_{ow}$ for naphthalene is 3.36, and its unitless Henry's law constant is 0.0174. Assuming everything is at equilibrium, please estimate the amount of naphthalene you would ingest if you were to eat only the fat blob.[3]

Strategy. Let us set up the mass balance equation and the partitioning equations and solve for the concentration in the fat:

$$M_{naphth} = C_{air}(1\,L) + C_{water}(1\,L) + C_{fat}(0.001\,L)$$
$$= 1\,mg$$
$$H' = \frac{C_{air}}{C_{water}} = 0.0174$$
$$K_{ow} = \frac{C_{fat}}{C_{water}} = 10^{3.36} = 2290$$
$$M_{naphth} = H'C_{water}(1\,L) + C_{water}(1\,L)$$
$$+ C_{fat}(0.001\,L) = 1\,mg$$

---

[3] From R.P. Schwarzenbach et al., p. 94; used with permission.

$$M_{\text{naphth}} = H'\left(\frac{C_{\text{fat}}}{K_{\text{ow}}}\right)(1\,\text{L}) + \left(\frac{C_{\text{fat}}}{K_{\text{ow}}}\right)(1\,\text{L})$$

$$+ C_{\text{fat}}(0.001\,\text{L}) = 1\,\text{mg}$$

$$C_{\text{fat}}\left(\frac{H'}{K_{\text{ow}}} + \frac{1}{K_{\text{ow}}} + \frac{0.001}{1}\right) = 1\,\text{mg/L}$$

$$C_{\text{fat}}\left(\frac{0.0174}{2290} + \frac{1}{2290} + \frac{0.001}{1}\right)$$

$$= 0.00144\,C_{\text{fat}} = 1\,\text{mg/L}$$

$$C_{\text{fat}} = 696\,\text{mg/L}$$

Naphthalene eaten = mass in $0.001\,\text{L}$

$$= \left(\frac{696\,\text{mg}}{\text{L}}\right) \times 0.001\,\text{L} = 0.7\,\text{mg}$$

5.8 WATER–AIR TRANSFER

As an example of a nonequilibrium partitioning process, let us work out the flux of a compound between the air over a lake and the water in the lake. This is an area of research that has been well studied, particularly for the transfer of PCBs into and out of the Great Lakes. Let us assume that the concentration of an organic compound in the air over the lake is C_{air} and the concentration in the water of the lake is C_{water}. The flux between the water and the air is

$$Flux = v_{\text{tot}}\left(C_{\text{water}} - \frac{C_{\text{air}}}{H'}\right)$$

Note that this equation uses the unitless value of the Henry's law constant. If everything is at equilibrium, then there is no flow across the interface $(F = 0)$, and thus, $C_{water} = C_{air}/H'$, which is the definition of the Henry's law constant. By convention, if the flux is negative, the flow is from the atmosphere into the lake, and if the flux is positive, the flow is from the lake into the atmosphere.

v_{tot} in the above equation is a mass transfer velocity across the air–water interface. It has units of velocity (usually cm/s), and it is made up of two parts: (a) the velocity of the compound through the boundary layer in the water to the interface (denoted by v_w) and (b) the velocity of the compound through the boundary layer in the air as it leaves the air–water interface (denoted by v_a). The total mass transfer velocity is given by

$$\frac{1}{v_{tot}} = \frac{1}{v_w} + \frac{1}{v_a H'}$$

For a given compound, the values of both v_w and v_a depend on wind speed over the water. The faster the wind, the faster the mass transfer takes place. The wind speed is usually denoted by u, and it has units of m/s. The value of v_a is derived from the velocity of water through air and is given by

$$v_a = (0.2\,u + 0.3)\left(\frac{18}{MW}\right)^{0.5}$$

where MW is the molecular weight of the compound of interest. The resulting units of v_a are cm/s.

The value of v_w is derived from the velocity of oxygen through water and is given by

$$v_w = 4 \times 10^{-4}\left(0.1\,u^2 + 1\right)\left(\frac{32}{MW}\right)^{0.5}$$

where u is in m/s and MW is the molecular weight. The resulting units of v_w are cm/s.

Consider *para*-dichlorobenzene (DCB), which is used as a toilet disinfectant. The following data are available for DCB: the molecular weight is 146 g/mol, the liquid vapor pressure is 1.3 Torr, and the saturated water solubility is 5.3×10^{-4} mol/L. In Lake Zurich, the measured concentration of DCB is 10 ng/L and the average wind speed is 2.3 m/s. What is this compound's flux into or out of Lake Zurich?

Strategy. We first calculate a unitless Henry's law constant:

$$H = \frac{P_L}{C_w^{sat}} = \left(\frac{1.3\,\text{Torr}}{1}\right)\left(\frac{1\,\text{atm}}{760\,\text{Torr}}\right)\left(\frac{\text{L}}{5.3 \times 10^{-4}\,\text{mol}}\right)$$

$$= 3.23\,\text{L atm/mol}$$

$$H' = \frac{H}{RT} = \left(\frac{3.23\,\text{atm L}}{\text{mol}}\right)\left(\frac{\text{mol K}}{0.082\,\text{L atm}}\right)\left(\frac{1}{288\,\text{K}}\right)$$

$$= 0.14$$

Now we need the air-side and the water-side mass transfer velocities:

$$v_a = (0.2 \times 2.3 + 0.3)\left(\frac{18}{146}\right)^{0.5} = 0.27\,\text{cm/s}$$

$$v_w = 4 \times 10^{-4}(0.1 \times 2.3^2 + 1)\left(\frac{32}{146}\right)^{0.5}$$

$$= 2.86 \times 10^{-4}\,\text{cm/s}$$

The overall mass transfer velocity is

$$\frac{1}{v_{tot}} = \frac{1}{2.86 \times 10^{-4}} + \frac{1}{0.27 \times 0.14}$$

$$v_{tot} = \left(\frac{1}{3492 + 26.5}\right) = 2.84 \times 10^{-4}\,\text{cm/s}$$

Note that the water transfer term (3492 cm/s) is much larger than the air transfer term (26.5 cm/s). This suggests that the transfer limit of this compound from water to air is the diffusion through the boundary layer of the water and not its removal from the surface by the air.

Assuming that the air concentration of DCB above Lake Zurich is very low such that we can call it zero, the flux out of the lake is simply

$$Flux = v_{tot}C_w = \left(\frac{2.84 \times 10^{-4}\,\text{cm}}{s}\right)\left(\frac{10\,\text{ng}}{L}\right)$$

$$\times \left(\frac{L}{10^3\,\text{cm}^3}\right)\left(\frac{60 \times 60\,\text{s}}{h}\right)\left(\frac{10^4\,\text{cm}^2}{m^2}\right)$$

$$= 102\,\text{ng}\,\text{m}^{-2}\,\text{h}^{-1}$$

Let us check this result with some data measured at Lake Zurich. This lake has an area of 68 km² and average depth of 50 m. There are two sources of DCB to the lake: sewage, which delivers 62 kg/ year, and flow from upstream, which delivers 25 kg/year. The downstream flow removes 27 kg/year from the lake. There is no accumulation of DCB in the lake's sediment. What is the evaporative flux of

DCB from this lake (in ng m^{-2} h^{-1}), and how does it compare to the above calculation?

Strategy. We can calculate the net loss of DCB from the lake as inputs minus outputs and assume that this is all lost by evaporation, which is a good assumption.

$$\text{Flux} = \frac{\text{Flow}}{\text{Area}} = \left[\frac{(62 + 24 - 27)\,\text{kg}}{\text{year}}\right]\left(\frac{1}{68\,\text{km}^2}\right)$$
$$= \left(\frac{59\,\text{kg}}{\text{year}}\right)\left(\frac{1}{68\,\text{km}^2}\right)\left(\frac{10^{12}\,\text{ng}}{\text{kg}}\right)$$
$$\times \left(\frac{\text{km}^2}{10^6\,\text{m}^2}\right)\left(\frac{\text{year}}{24 \times 365\,\text{h}}\right) = 99\,\text{ng m}^{-2}\,\text{year}^{-1}$$

This agrees with the previous calculation; in fact, the rather exact agreement is certainly just dumb luck.

The evaporative half-life of a compound in a lake can be derived by noting

$$t_{1/2} = \frac{\ln(2)}{k} = \ln(2)\tau$$

The evaporative residence time is given by

$$\tau = \frac{M}{\text{Flow}} = \frac{C_{\text{w}}V}{\text{Flow}}$$

Remembering that

$$\text{Flow} = \text{Flux} \times A$$

and

$$V = A\bar{d}$$

we can substitute these in the above equation and get

$$t_{1/2} = \ln(2) \frac{C_w A \bar{d}}{\text{Flux} \times A}$$

In the case of evaporation, we know that

$$\text{Flux} = C_w \times v_{tot}$$

Hence, the evaporative half-life is

$$t_{1/2} = \frac{(\ln 2) C_w A \bar{d}}{C_w v_{tot} A} = \frac{(\ln 2) \bar{d}}{v_{tot}}$$

In the case of DCB in Lake Zurich,

$$t_{1/2} = \left(\frac{\ln(2)\,s}{2.84 \times 10^{-4}\,\text{cm}}\right)\left(\frac{50\,\text{m}}{1}\right)\left(\frac{100\,\text{cm}}{\text{m}}\right)$$
$$\times \left(\frac{\text{day}}{60 \times 60 \times 24\,\text{s}}\right) = 140\,\text{days}$$

5.9   PROBLEM SET

1. Estimate the maximum concentration of 1,2,4-tri-chlorobenzene (1,2,4-TCB) in rainbow trout swimming in water with a 1,2,4-TCB concentration of 2.3 ppb. Please give your answer in ppm. The molecular weight of this compound is 181.5 g/mol, and its $\log K_{ow}$ (at 25°C) is 4.04.

2. The triazines are an important class of herbicides; see the structure below. In a certain class, R_1 to R_4 vary as follows:

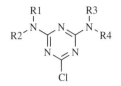

Compound	R_1	R_2	R_3	R_4
1	C_2H_5	C_2H_5	C_2H_5	C_2H_5
2	C_3H_7	H	C_2H_5	C_2H_5
3	C_2H_5	H	C_2H_5	C_2H_5
4	C_3H_7	H	C_3H_7	H
5	C_2H_5	H	C_3H_7	H
6	C_2H_5	H	C_2H_5	H

where C_2H_5 is an ethyl and C_3H_7 is an isopropyl group. The log K_{ow} value for compound 5 is 2.64. What are the log K_{ow} values of the other compounds?

3. Calculate the flux of 1,1,1-trichloroethane (TCE) between the air and the surface water of the Arctic Ocean at a temperature of 0°C and an average wind speed of 10 m/s. Use the following data: $C_a = 0.93\,\text{ng/L}$; $C_w = 2.5\,\text{ng/L}$; $H = 6.5\,\text{atm L/mol}$ at 0°C; molecular weight $= 133.4\,\text{g/mol}$.

4. Due to an accidental spill, a significant amount of TCE (see just above) has been introduced into a small, well-mixed pond (volume $V = 1 \times 10^4\,\text{m}^3$, total surface area $A = 5 \times 10^3\,\text{m}^2$, $T = 15°C$). Measurements carried out after the spill, during a period of 1 week, showed that TCE was eliminated from the pond by a first-order process with a half-life of 40 h. Because export of TCE by the outflow of the pond can be neglected, and because it can be assumed that

neither sedimentation nor transformations are important processes for TCE, the observed elimination has to be attributed to exchange to the atmosphere. (a) Calculate the average v_{tot} of TCE during the time period considered by assuming that the concentration of TCE in the air above the pond is very small, that is, $C_w \gg C_a/H'$. (b) What is the average wind speed that corresponds to this v_{tot} value?

5. Lake William has an area of 120 km^2 and an average depth of 62 m. The average concentration of tetrachloromergetene (TCM) is 40 ng/L; its water solubility is 210 ppm; its molecular weight is 236; and its vapor pressure is 12 Torr. Assume a wind speed of 5 m/s, a temperature of 300 K, and a very low air concentration. (a) What are the Henry's law constant, the overall exchange velocity (v_{tot}), and the evaporative flux of TCM from Lake William? What is the evaporative half-life of TCM in this lake? (b) Lake William has only one input: River Arthur flowing at $25 \times 10^9 \text{ m}^3/\text{year}$. River Lancelot is the only output. The average concentration of TCM in River Arthur is 75 ng/L, and in River Lancelot it is 50 ng/L. Please calculate the evaporative flux of TCM from this lake and compare it to the predicted value calculated above.

6. Lake Harry has an area of 180 km^2, an average depth of 50 m, and a water residence time of 0.3 year. The flux of a pollutant to the sediment is $10 \text{ ng cm}^{-2} \text{ yr}^{-1}$, and its concentration in the water averages 1.8 ng/L. Assume v_{tot} is 1 cm/h. Calculate (a) the water flow rate in and out of the lake and (b) the pollutant's evaporative flux.

7. The evaporative half-life of toluene from Lake Hal is 120 days. What is the unitless Henry's law constant for this compound? The molecular weight of toluene is 92, and the average depth of Lake Hal is 75 m. Assume a wind speed of 5 m/s.

8. Lake William (10^4 km^2 in area) receives PCBs from the Richard River and rainfall and loses PCBs only by sedimentation. The river concentration is 1.0 ng/L, and its flow rate is 10^4 m^3/s. The rain concentration is 20 ng/L, and its delivery rate is 80 cm/year. The PCB concentration in the lake's outlets is negligible. What is the PCB flux to the lake's sediment?

9. The suspended solids in Lake Alfred have 2% organic matter. The aquatic (dissolved) concentration of tetrabromobadstuff is 60 ng/L, and its water solubility is 2.8 μmol/L. What is the concentration of this compound on the suspended solids?

10. The K_{ow} values of PCBs vary with the number of chlorines. Imagine that two PCBs are released into a lake in a 1:1 ratio; the first has three chlorines and the second has five chlorines. What would the ratio of these two PCB concentrations be in fish?

11. The concentration of tetrachlorocrudene in trout in Lake Charlotte averages 0.2 ppm; in this lake's water, the concentration is 7 ng/L. What is this compound's octanol–water partition coefficient?

12. From annual measurements, it has been determined that the residence time of PCBs in Lake Charlotte is 3.4 years. Assume that the only input is from rainfall and that input equals output. What is the concentration of PCBs in this lake? The following data

may be helpful: average depth of the lake $= 50\,m$, and PCB concentration in rain $= 40\,ppt$.

13. Assume Lake Ontario is fed only by the Niagara River (flow rate $= 7500\,m^3/s$) and drained only by the St Lawrence River. The concentration of penta-fluoroyuckene (a very nonvolatile compound) is 2.7 ppt in the Niagara and 1.2 ppt in the St Lawrence. What is this compound's average flux to the lake's sediment? Lake Ontario has an area of 20,000 km^2.

14. What is the steady-state body burden of heptachlor-obiphenyl in a 1.0-kg lake trout taken from water containing 0.09 ppt of heptachlorobiphenyl? Please assume a log K_{ow} for this compound of 6.5.

15. The concentration of a particular PCB congener in the water of northern Lake Michigan is 0.15 ng/L. Assume that the water solubility of this PCB congener is 0.01 μmol/L, that its vapor pressure is 5×10^{-9} atm, and that its molecular weight is 320. What is the water to air flow rate (in tonnes per year) of this compound on a warm summer day in northern Lake Michigan? Assume a water temperature of 25°C and a wind speed of 5 m/s. The area of northern Lake Michigan is $2 \times 10^4\,km^2$.

16. Trout in Lake Huron are an important game fish. Please assume the following: The concentration of hexachlorobonserene (HCB) in water from Lake Huron is 15 ppb, the half-life of HCB in trout is 4 days, and HCB's log K_{ow} value is 5.43. (a) What is the steady-state concentration of HCB in Lake Huron trout in ppm? (b) Relatively clean trout are introduced into Lake Huron. How long (in days) after this event would it take for the concentration

of HCB in these trout to reach 75 ppm? (c) Clean trout are put in Lake Huron water for 1 week, then transferred to clean water for another week. What is the concentration (in ppm) of HCB in these fish at the end of this time?

17. In a simple model of tetrabromoxylene (TBX) flow into and out of Lake Ontario (volume = $1.67 \times 10^{12} \, \text{m}^3$, average depth = 85.6 m), there are three inputs and three outputs. Unfortunately, only three inputs and two outputs have been characterized. TBX enters the lake through rain (concentration = 10 ng/L), rivers (344 kg/year), and streams and creeks (102 kg/year). TBX leaves the lake through rivers (310 g/day), volatilization from the lake surface (251 kg/year), and sedimentation. Assume that TBX is at steady state in the lake and that its residence time is 7.2 years. (a) What is the flow rate of TBX to the sediment? (b) What is the concentration of TBX in the lake water? (c) Fifteen fish were captured, and their average TBX concentration was determined to be 0.21 ppm. What is the biota partitioning coefficient? Please give this answer as $\log K_B$.

18. A research vessel on Lake Ontario had a slow leak of diesel fuel as it cruised around the lake sampling sediment. Over a few days, this vessel lost 50 gal of diesel fuel. The research scientists on this vessel patched the leak, and then, being a conscientious group, they began to wonder about the environmental ramifications of this spill. They assumed that diesel fuel was composed entirely of n-hexadecane ($C_{16}H_{34}$), and looking

through a reference book they had brought with them (they were *very* conscientious), they found the following information about n-hexadecane: MW $= 226.4$; density $= 0.773\,\mathrm{g/mL}$; $-\log C_w^{sat} = 7.80\,\mathrm{mol/L}$; $-\log P_V = 5.73\,\mathrm{atm}$. Assume that the n-hexadecane was mixed uniformly throughout the lake, that evaporation is the only loss mechanism, and that the wind speed averaged $4\,\mathrm{m/s}$. How long would it take for the spill to dissipate? Assume five half-lives are sufficient for the spill to dissipate.

CHAPTER 6

TOXIC ENVIRONMENTAL COMPOUNDS

One of the most important innovations in agriculture and public health was the development and application of compounds designed to control (kill) insects that interfere with the efficient growing of numerous crops and that transmit diseases to people. At first, these compounds were inorganic compounds that killed more or less everything in sight, but in the 1930s and 1940s, organic compounds (DDT being the most famous example) came on the market. These compounds were designed to specifically target insects rather than mammals. Eventually, compounds were developed that also targeted weeds, thus allowing the farmer to increase the yield of a crop per unit area of land.

Despite the social and economic benefits of these compounds, problems developed because some of these compounds, once released, did not degrade in the environment. Because of their environmental persistence, some of these compounds have had unintended consequences; for example, DDT caused egg shell thinning and thus affected the reproduction of certain types of birds. This problem was brought to the public's attention by the famous book *Silent Spring*.[1] Perhaps as a result, many of these early pesticides are no longer on the

---

[1] R. Carson, *Silent Spring*, Houghton Mifflin, Boston, 1962.

Elements of Environmental Chemistry, by Ronald A. Hites
Copyright © 2007 John Wiley & Sons, Inc.

market and have been replaced by compounds that are less environmentally persistent (but somewhat more toxic to mammals). However, just because they are no longer on the market, this does not mean that they have disappeared as environmental problems. In fact, because of their persistence, large amounts of many of these so-called legacy pesticides are still with us and still show up in our food. The global community has taken notice of the problem and has established the Stockholm Convention aimed at eliminating the environmental release of several of these persistent pesticides.

The Stockholm Convention is described as follows on the United States Environmental Protection Agency's web site: "The United Nations Environment Program sponsored negotiations to address the global problems of POPs (persistent organic pollutants). The United States joined forces with 90 other countries and the European Community to sign a groundbreaking United Nations treaty in Stockholm, Sweden, in May 2001. Under the treaty, known as the Stockholm Convention or the POPs treaty, countries agree to reduce or eliminate the production, use, and/or release of 12 key POPs. The Convention specifies a scientific and procedural review mechanism that could lead to the addition of other POPs chemicals of global concern."

"The global POPs agreement initially covers the 'dirty dozen,' which includes nine pesticides (aldrin, chlordane, DDT, dieldrin, endrin, heptachlor, hexachlorobenzene, mirex, and toxaphene), and three industrial chemicals (PCBs) and unintentional by-products (dioxins and furans) of industrial and combustion processes. The purpose of the agreement is to reduce releases of POPs chemicals on a global basis. The agreement requires all Parties to stop production and new uses of intentionally produced POPs, with limited exceptions, and to implement strong controls on sources of by-product POPs to reduce emissions. In addition, a strong financial and technical assistance provision in the agreement will provide support to developing countries enabling them to implement the pollution reduction requirements of the treaty." The United States has signed this treaty, but it is not yet an "official partner."

This chapter takes us through the names, structures, and stories of some of these legacy pollutants and of some of their replacements that are now in common use. But first we need a bit of an overview.

6.1 PESTICIDES

Pesticides include many types of chemicals that are spread around in the environment to kill some specific sort of pest, usually insects (insecticides), weeds (herbicides), or fungi (fungicides).[2] The total worldwide use of pesticides is now about a million tonnes per year. Historically, insecticides have gone through several "generations," which we will categorize as follows:

Generation 0. These include physical methods of pest control such as rocks, pieces of wood, shoes, fly paper,

---

[2] A note for the Latin scholars among you: In these words, what does the "-icide" suffix mean? What would fratricide mean?

fly swatters, and so on. These methods are environmentally friendly, but not very effective if you are up to your armpits in locusts.

Generation I. These were inorganic compounds such as Paris Green [$Cu(AsO_3)_2$] and Green Lead [$Pb_3(AsO_4)_2$]. These compounds did not kill just insects, but they also killed mammals (including the occasional farmer's child). They are not used any more.

Generation IIa. These were the now infamous chlorinated organic compounds such as DDT. These compounds have substantial ecological problems and are largely banned in the industrialized world. However, most of these compounds have such long environmental lifetimes that they can still be found (sometimes in surprisingly high amounts) in current environmental samples. Most of these compounds were substantially more toxic to insects compared to mammals.

Generation IIb. After the problems of the chlorinated organics became clear, the chemical and agricultural industry went to organophosphorous-based compounds. Although these compounds are not particularly long lived in the environment, they are more toxic to mammals as compared to the chlorinated compounds.

Generation III. These compounds often mimic some feature of the insect's natural hormones. These include pheromones, insect growth regulators, and other compounds that simulate these natural products.

The following sections present the structures of representative pesticides. These compounds are either historically or commercially important.

6.1.1 Diphenylmethane Analogs

DDT (Dichlorodiphenyltrichloroethane)

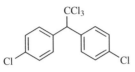

Note that this is the *para para* (abbreviated as p,p') isomer. The term *para* means that the chlorine atoms are positioned on the benzene ring across from the linkage to the rest of the molecule. DDT won Paul Muller the Nobel Prize in medicine in 1948 for malaria control because DDT killed the mosquitoes that transmitted this disease. At least 2 billion kilograms of DDT has been used worldwide since about 1940. Because of problems with calcium metabolism in birds (egg shell thinning), DDT has been banned in most industrialized countries since about 1970–1975. By the loss of HCl, p,p'-DDT degrades relatively quickly to DDE (see below), which is almost permanently stable in the environment. DDT, however, is still produced and used to control malaria in developing countries because it is cheap to manufacture and simple to use.

DDE

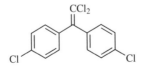

Methoxychlor

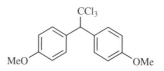

Methoxychlor is structurally related to DDT, but it is much less persistent and is, therefore, still in use. For example, it can be found in household quality flea dip.

6.1.2 Hexachlorocyclohexanes

These are usually abbreviated as HCHs for obvious reasons. The most well known of these compounds is lindane, which is also known as the *gamma* isomer of HCH or γ-HCH. Its structure is given below.

Lindane (or γ-HCH)

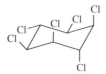

Note the orientation of the chlorines relative to the cyclohexane ring; this is the *aaaeee*-isomer, where *a* is axial and *e* is equatorial. HCHs are abundant in the Love Canal dump because the Hooker Chemical Corp. in Niagara Falls, NY, made them. HCHs have environmental half-lives of a few years, and they have been found in mammals from the Arctic. Lindane's use in the United States is being phased out starting in 2009.

6.1.3 Hexachlorocyclopentadienes

These compounds are all based on the Diels–Alder reaction of hexachlorocyclopentadiene (also called C-56) with other compounds such as cyclopentadiene.

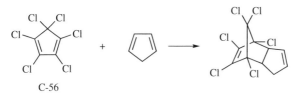

C-56

In this case, the resulting compound is called chlordene, but it is *not* a pesticide. Rather, it is used to make other pesticides such as chlordane and heptachlor.

Chlordane

Chlordane was widely used as a termiticide in and around homes, but it has been banned in most developed countries since about 1985. It has an environmental half-life of about 5 years. It degrades to oxychlordane (see below), which is very stable in the environment.

Oxychlordane

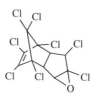

Heptachlor

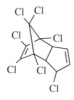

Heptachlor was a widely used pesticide, but it is now banned. It is closely related to chlordane, and like chlordane, heptachlor degrades quickly in the environment to its epoxide, in this case, called heptachlor epoxide (see below). In 1981–1987, this latter compound was a problem in milk from cows in Oahu, Hawaii, because of heptachlor contamination of pineapple greens, which had been used as cow food.

Heptachlor epoxide

Two other compounds that were made from C-56 are mirex and kepone. Both were widely used to kill ants, particularly fire ants in the southern part of the United States. Think of these two compounds as two molecules of C-56 stacking up on top of one another.

Mirex

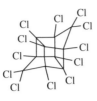

Mirex is widely distributed in the sediment of Lake Ontario because of two sources: (a) Hooker Chemicals (now called OxyChem) in the city of Niagara Falls, New York; and (b) Armstrong Cork in the city of Oswego, New York. Mirex is now banned.

Kepone

Kepone is simply a version of mirex in which a CCl_2 group has been oxidized to a carbonyl group $(C=O)$. Kepone contaminates the sediment of the James River near Hopewell, VA, due to a manufacturing problem. It is now banned.

6.1.4 Phosphorous-Containing Insecticides

6.1.4.1 Phosphates

Dichlorvos

$$(CH_3O)_2\underset{\underset{O}{\|}}{P}OCH{=}CCl_2$$

This is a relatively volatile compound that was used to kill flies indoors. It was in the famous yellow plastic "Shell No-Pest Strips." This compound is no longer in use.

Glyphosate (Round-Up)[3]

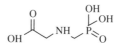

Although we will not get to herbicides until later, it is appropriate to mention a very popular herbicide here. Glyphosate (also widely known as Round-Up) is the most widely used pesticide in the United States. In 2001, 40–50 million kilograms of glyphosate was used in the United States. This herbicide is used in conjunction with "Round-Up Ready" corn and soybeans, crops that have been genetically modified to be resistant to Round-Up. Thus, the entire field can be sprayed, and the weeds will die but the crop will not.

6.1.4.2 Phosphorothioates

Parathion

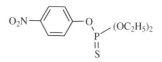

Like most of the phosphorous-based insecticides, parathion is effective for insect control, but it also has

---

[3] Some of these compounds have both generic and trade names. The trade names are the specific property of a company. In these cases, the generic name is given followed by the trade name in parentheses.

marked mammalian toxicity. In this case, the LD_{50} for rats is about 3 mg/kg.[4] Parathion has been used since about 1950, but it is off the market now.

Chlorpyrifos (Dursban)

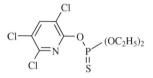

This is one of the most popular insecticides in use today. In 2001, it ranked second in popularity among the insecticides (right after malathion, see below); 4–5 million kilograms were used in the United States. It has replaced chlordane for termite control and is used widely indoors for cockroach control. The use of chlorpyrifos is now being phased out, and it is not clear what will replace it.

6.1.4.3 Phosphorodithioates

Malathion

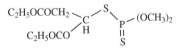

Malathion is probably the least toxic of the phosphorous-based insecticides; nevertheless, it is now the most widely used insecticide in the United States; in

---

[4] LD-50 is the dose that it would take to kill 50% of a population.

2001, 10–15 million kilograms was used. Malathion has been sprayed in urban areas (such as Los Angeles) and used in flea dip for dogs.

6.1.5 Carbamates

Carbamates all have the general structure of $RO(CO)NHCH_3$. It turns out that only carbamates with a monomethyl-substituted amine group are biologically active. Although these compounds are not widely used any more, there are two carbamates, the structures of which we should be aware.

Carbaryl

Carbaryl has been around a long time and was approved for home use. The bicyclic aromatic ring system is called naphthalene. Learn the correct spelling of this moiety.

Carbofuran (Furadan)

This compound is now being banned.

6.1.6 Natural Product Simulants

Methoprene

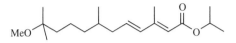

Methoprene simulates a mosquito growth-regulating hormone. When the mosquito larvae are exposed to this compound, the insects fail to change to an adult. This compound is specific to mosquitoes and is non-toxic and environmentally friendly.

Permethrin

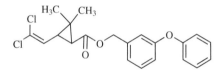

Note the cyclopropyl ring with the two methyl groups on one carbon—this is unusual. Permethrin is an anthropogenic compound that simulates the insecticidal properties of pyrethroids found in chrysanthemums. It is very common in household products such as Raid. It is assumed to be nontoxic to mammals.

6.1.7 Phenoxyacetic Acids

245T

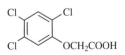

245T is a herbicide generally aimed at killing broad-leaf weeds. 245T came to the public's attention because it was a component of Agent Orange, a herbicide used by the U.S. military in Vietnam in the 1960s and 1970s. Unfortunately, this 245T included dioxin (more on this below) as an impurity.

Silvex

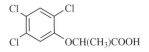

As its name indicates, silvex has been widely used on trees. 245T and silvex are banned in most industrialized countries because of dioxin impurities.

245T and silvex are both produced from 2,4,5-trichlorophenol, which has a tendency to dimerize to give 2,3,7,8-tetrachlorodibenzo-*p*-dioxin (2378-TCDD, see below). This compound is known far and wide as "the most toxic man-made chemical" (at least it is very acutely toxic to male guinea pigs). Dioxins are not pesticides, but they have been present in pesticides because of manufacturing impurities. There have also been accidental releases of dioxins during the course of pesticide and chlorinated phenol manufacturing; the most famous of which was in Seveso, Italy. Dioxins have also been distributed in the environment by the illicit dumping of hazardous waste; the most famous of which was in Times Beach, Missouri.

2378-TCDD (this compound is not a pesticide)

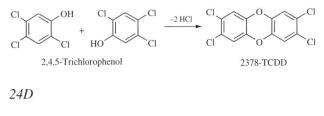

2,4,5-Trichlorophenol   2378-TCDD

24D

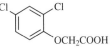

Unlike 245T and silvex, 24D was not produced from 2,4,5-trichlorophenol and therefore, it never had 2378-TCDD as an impurity. As a result, 24D is still on the market. In fact, it is the fifth most widely used pesticide in the United States; in 2001, 15–20 million kilograms was used.

6.1.8 Nitroanilines

Trifluralin (Treflan)

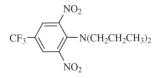

Trifluralin is a widely used herbicide, ranking 12th in popularity. Between 5 and 10 million kilograms was used in 2001 in the United States. Treflan was manufactured in Indiana by Ciba-Geigy since 1963, and in fact, fish near this manufacturing site (on the Wabash River) had been contaminated with this compound, resulting in somewhat yellow fish tissue.

Pendimethaline (Prowl)

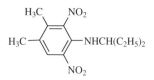

This is another popular dinitroaniline herbicide, ranking 11th in usage in the United States in 2001 at 5–10 million kilograms.

6.1.9 Triazines

Atrazine

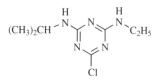

The triazines are very widely used herbicides. The best known and the most widely used of these compounds is atrazine. This compound ranks second in terms of usage in the United States (just behind glyphosate); 35–40 million kilograms of atrazine was used in 2001. This compound has been manufactured by Ciba-Geigy since

1958, and it must be one of their most successful products. Atrazine is used in great abundance on corn. Because of its widespread use, atrazine has shown up in drinking water downstream of agricultural areas; for example atrazine is present in the drinking water of New Orleans. In fact, the United States Environmental Protection Agency has regulated atrazine levels to be less than a few ppb in drinking water throughout the United States. Luckily, the triazines are not particularly toxic to mammals.

6.1.10 Acetanilides

All of these compounds feature a $ClCH_2CON-$ functional group, and all are very popular herbicides.

Acetochlor

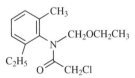

This compound was the fourth most used pesticide in the United States in 2001 at 15–20 million kilograms.

Alachlor (Lasso)

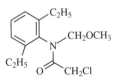

This was the 16th most popular pesticide in the United States in 2001 at 3–5 million kilograms usage, but in 1987, it was the second most popular pesticide. This demonstrates that the pesticide market is always changing as new compounds are developed and as older compounds go out of patent protection.

Metolachlor (Dual)

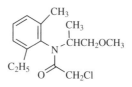

This is also a popular herbicide, ranking 9th in 2001 in the United States with a usage of 15–25 million kilograms.

6.1.11 Fungicides

These compounds are used to prevent fungi from destroying building materials, especially wood. Several are based on the pentachlorobenzene structure.

When R = OH, the compound is called pentachlorophenol (PCP); this has probably been the most widely used wood preservative, but it is no longer used because of dioxin contamination in the technical grade product. When R = Cl, the compound is called hexachlorobenzene; this is also a popular fungicide used especially on seeds. The use of hexachlorobenzene is banned as a fungicide, but it is still used as a synthetic intermediate for the small-scale production of other chemicals.

6.2 MERCURY

Mercury (Hg) is a toxic metal—it is one of the so-called "heavy metals." Hg is a neurotoxin that causes damage to the central nervous system (CNS), which consists of the brain and associated parts. Hg is active at about $50 \, \mu g/100 \, mL$ of blood (500 ppb). Central nervous system damage manifests itself as quarrelsome behavior, headaches, depression, and muscle tremors. The classic example of mercury poisoning is the mad hatter, caused by mercury exposure during the felt-making process.

The toxicity of mercury depends on its exact form (and valence state):

- Hg^0 inert as a liquid, but the vapor is toxic;
- Hg_2^{2+} or Hg(I)   not very soluble, and it is not very toxic;
- Hg^{2+} or Hg(II)   not toxic, but it is easily methylated (see below);
- CH_3Hg very toxic and lipophilic; it easily enters the CNS;
- CH_3HgCH_3 not toxic, but it is easily transformed to CH_3Hg.

Modern exposure to mercury is seen in miners of cinnabar (HgS), in workers in the chlor-alkali industry, in dental workers, and in battery manufacturing. Almost all of these exposures are now under control.

Losses of mercury to the environment had been from the chlor-alkali industry (20–30 t/year in the United States), coal burning (about 3–5 t/year), and oil burning (15–25 t/year). The chlor-alkali losses have been considerably reduced in recent years by the simple expedient of not using mercury in that industry.

What is the chlor-alkali industry? It is the production of sodium hydroxide (NaOH) and chlorine gas (Cl$_2$) by the electrolysis of sodium chloride (NaCl) in a water solution. In the old days, mercury was used as one of the electrodes; now titanium is used. Figure 6.1 shows a

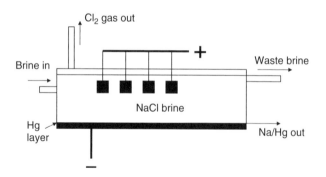

Figure 6.1 Schematic of a chlor-alkali cell in which an electric current is passed through a sodium chloride solution (brine). Chlorine gas is produced at the anode (note the paddle-like electrodes with the positive charge at the top), and sodium dissolves in the cathode (note the negative charge at the bottom). In this case, the cathode is a pool of elemental mercury, and the sodium amalgam is later hydrolyzed to produce sodium hydroxide.

flowing mercury cell for the electrolytic production of NaOH and chlorine gas.

Mercury poisoning in Minamata Bay, Japan, caused a serious environmental disaster. This is what happened:

- Like most of Japan, the people in this small town ate lots of fish (about 350 g/day).
- The town was the home of a plant that converted acetylene into acetaldehyde; mercury was used as a part of the process.
- During the period 1932–1968, about 400 t of mercury had been dumped into the bay.
- By 1953, it was noticed that cats were suffering from CNS damage.
- By 1960, about 1300 people were showing CNS damage.
- Both the feline and human effects were worse near the plant.
- This biological effect was traced to "heavy metals" in the fish that both the cats and the people ate.
- It was not clear at that time how a metal could bioaccumulate in fish. At that time, mercury was thought to be relatively non-lipophilic.
- Eventually, it was learned that mercury is converted by bacteria in anoxic sediments into methyl mercury (CH_3Hg), which is lipophilic.
- By 1960, cat-feeding studies had linked the CNS effects to CH_3Hg.
- The concentrations of CH_3Hg in the sediments near the plant were measured to be 10–2000 ppm, and in the fish from the bay, the concentrations were

5–40 ppm. Mercury is now regulated at a level of 0.5 ppm.

- By 1968, the discharge was stopped, and about 200 people were dead.
- The contaminated sediment is still in place. Why?

The problem was the lack of knowledge that mercury could be converted from a nonlipophilic form into a lipophilic form in situ in the sediment. Most industrial sources of mercury are now under control except for municipal solid waste incinerators, in which the main problem seems to be mercury in batteries.

6.3 LEAD

Like mercury, lead (Pb) is also a "heavy metal" that causes CNS damage. At lower doses, it can cause anemia and kidney damage. The dosimetry is based on blood levels:

- $<10\ \mu g/100\ mL$[5]   normal ambient average;
- $40\ \mu g/100\ mL$ behavior and IQ effects;
- $70\ \mu g/100\ mL$ peripheral neuropathy;
- $>190\ \mu g/100\ mL$ confusion and convulsions.

In a groundbreaking paper, Needleman et al. showed significant effects of lead on children's learning

---

[5] In some cases the 100 mL unit here is abbreviated as "dL" for deciliter. Physicians are particularly fond of this notation, so watch out for it.

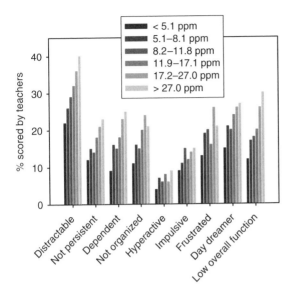

Figure 6.2 Distribution of negative ratings by teachers on nine classroom behaviors in relation to tooth lead concentrations. [Replotted from H.L. Needleman et al., *New England Journal of Medicine*, **300**, 689–695 (1979).]

behavior and on their IQ. Figure 6.2 shows the distributions of negative ratings by teachers on nine classroom behaviors in relation to "baby teeth" lead concentrations. The average IQ for the low-lead group was 106.6, and for the high-lead group, it was 102.1. This is a difference of 4.5 IQ units, which was significant with 97% confidence.

Historically, lead has had a variety of uses:

- pottery glaze (since at least 6000 B.C.);
- cooking utensils (may have caused the Fall of Rome);

- plumbing (now rare, except in New England);
- solder (e.g., sealing canned milk);
- storage batteries in cars (largest current use in the United States);
- pigments in paint (a problem with chipping paint in older dwellings) [the older pigment was called white lead, $Pb(OH)_2 \cdot 2PbCO_3$; it has now been replaced by TiO_2];
- gasoline additive [$Pb(C_2H_5)_4$] was used to prevent knocking in engines; it is now largely banned in the developed world. In 1980, about 10^5 tonnes of this compound was produced in the United States. Blood levels of lead in the United States have generally tracked the use of this compound and are now decreasing.

Lead's total worldwide use is now about 3×10^6 t/year, and most of that is in the storage battery business.

The environmental effects of lead had been almost all from auto exhaust, but these effects have decreased considerably since 1973, when unleaded gasoline became available. Gasoline in the United States now has almost no lead in it, but on a global scale, about 10^5 t/year of lead enters the atmosphere, mostly from the use of leaded gasoline. The lead concentrations in the atmosphere are now about 0.1 $\mu g/m^3$ in rural areas and up to about 0.6 $\mu g/m^3$ in urban areas. The human dietary intake is now about 0.2 mg/day in the United States, which is not a problem.

The regulatory approach in the United States has focused on the elimination of lead from gasoline. In 1975, gasoline contained about 4 g Pb/gal; now it is almost zero. The cost of achieving this reduction has been about 600 M$, but the cost saving in healthcare has been about 1800 M$. The goal was to get human blood levels to <3 µg/100 mL; the current level is about 2.8 µg/100 mL. The regulations seem to have worked. Incidentally, unleaded gasoline also saves the catalytic converter (used for hydrocarbon and NO control).

Both mercury and lead have been environmental success stories about which the United States Environmental Protection Agency and the environmental community can feel proud.

6.4 PROBLEM SET

1. Give the name and structure of a pesticide described by the following phrases. Use a given compound only once.
   a. Won Paul Muller the Nobel Prize in 1948.
   b. A triazine-based herbicide with exactly five nitrogen atoms.
   c. Has exactly seven chlorine atoms.
   d. Found in the sediment of the James River, Virginia.
   e. A "natural" pesticide.
   f. Has at least two fluorine atoms.
   g. A dimer of C-56.
   h. Has exactly six chlorine atoms and is in Love Canal.
   i. Similar to DDT but has two oxygen atoms.
   j. Used in Vietnam by the U.S. Army.
   k. Used in Bloomington to kill cockroaches.

l. Is an excellent wood preservative.

m. Made from C-56.

n. Used with genetically modified seed.

o. Currently the most widely used insecticide.

p. Similar to a natural pesticide found in chrysanthemums.

2. The following series of questions all relate to a
   "one-compartment" toxicological model for the
   uptake of a toxic substance from water by an aquatic organism. First-order rate constants k_1 and k_2
   are, respectively, for uptake from the water and loss
   from the organism (metabolism and excretion back
   to the water). Numerical starting conditions for the
   problem are C_w (concentration of toxicant in
   water) = 0.020 ppm; half-life for clearance of the
   toxicant from the organism = 3 days; log
   K_{ow} = 5.48. (a) Calculate the steady-state concentration of toxicant in the organism. (b) How long
   would it take for the concentration of the toxicant in
   the clean organism to reach 5 ppm? Assume the
   water to be an infinite reservoir of the toxicant.
   (c) A minnow is placed in a large tank and left there
   for 4 days. It is then transferred to a large tank of
   clean water. What is the concentration of the toxicant in its tissues after the next 4 days?

3. The detection limit of many chlorinated pollutants,
   such as DDT, chlordane, and PCBs, is on the order
   of 5 pg introduced into to a gas chromatographic
   column. Assuming a sample of human adipose
   tissue contains 34 ppt of DDT, that the extraction
   procedure is 75% efficient, and that the equivalent
   of 5% of the final sample can be injected into the gas

chromatograph for one analysis, what is the minimum size (in cm^3) mass of human adipose tissue that must be removed from a volunteer in order to detect this amount of DDT?

4. In the "Times Beach" incident, a waste oil dealer removed about 20,000 gal of oil contaminated by ~30 ppm of 2,3,7,8-tetrachlorodibenzo-p-dioxin from a hexachlorophene manufacturing plant. (a) What mass of dioxin was involved? (b) Some of this oil was sprayed in horse arenas, and in some spots in these arenas, the soil dioxin concentration was about 2000 ppb. What mass of this soil needed to be ingested by a 100 g guinea pig to reach the LD_{50} of 0.6 µg/kg?

5. The World Health Organization sets a standard of <0.2 mg for each 70 kg person per week as an acceptable mercury intake. In Canada, fish from the Great Lakes are considered edible if their mercury content is below 0.5 ppm. Are these values compatible?

6. Assume the average worker in the auto tunnels into and out of Manhattan had a blood lead level of 155 µg/L and that 25 µg/day of this lead is either excreted or deposited to bone. What is lead's residence time in the blood of this worker?

7. A lead recycling plant begins operation on the shores of a hitherto clean lake of volume 3.0×10^6 m^3. It discharges into the lake 12 m^3/h of waste containing 15 ppm of lead. The other inflow and outflow of the lake are rivers with flow rates of 8400 m^3/h. (a) What is the steady-state concentration of lead in the lake? Assume the

lake is well mixed and that it has no other source or sink for lead. (b) What is the residence time of lead in the lake at the steady state? (c) How long does it take for the lead level to reach 99% of its steady-state value?

8. In Eagle Lake (in Arcadia National Park) about 1350 fish were found dead one day. The estimated fish population in this lake is about 2700. Rangers suspect that the cause of death was 2,4,5-trifluoroyuckol, which was found in the water at a concentration of 20 ppb and which has a lethal dose to 50% of the fish population (LD-50) of 25 mg/kg on a whole fish basis. What is the water solubility of this pollutant (in μmoles/L)? Assume a typical fish in this lake weighs 0.4 kg and has a lipid content of 14% by weight.

9. Recent measurements of the melt-water from high-altitude snow in Greenland and the Himalayas have given pH values of 5.15 instead of the expected value. Some of this difference can be explained by the variation of the equilibrium constants as a function of temperature. (a) Please calculate the pH of water at $0°C$ in equilibrium with CO_2 at its current partial pressure. (b) What would the partial pressure (in ppm) of CO_2 have to be in order to get precipitation of pH $= 5.15$ at this temperature? (c) What do you conclude? Note that at $0°C$ $pK_H = 1.11$, $pK_{a1} = 6.57$, and $pK_{a2} = 10.62$.

10. If Treflan had its two nitro $(-NO_2)$ groups removed and replaced with two methyl $(-CH_3)$ groups, by what factor (how many times) would the new compound be more (or less) lipophilic than Treflan?

11. Assume a chemical is fed to some animal on a regular basis (say once a day). An approximate expression for the average concentration of this chemical after at least 10 doses is

$$C = \frac{1.44 D_0 \, f t_{1/2}}{\tau}$$

where each dose D_0 is given at an interval of τ; the fraction of dose absorbed is f; and the compound has a half-life in the animal of $t_{1/2}$. Please derive this expression. *Hints*: You will find the following series expansions useful: (a) $1 + x + x^2 + x^3 + \cdots = 1/(1 - x)$ if $x < 1$ and (b) $e^x = 1 + x + x^2/2! + x^3/3! + \cdots$.

12. For the reaction

$$NO + O_3 \rightarrow NO_2 + O_2$$

the rate constant is 1.8×10^{-14} cm^3/s at 25°C. The concentration of NO in a relatively clean atmosphere is 0.10 ppb and that of O_3 is 15 ppb. At these concentrations, what is the rate of this reaction? Assuming an excess of ozone, what is NO's half-life due to this reaction?

13. Nitrogen oxide is formed in the dark by the dissociation of dinitrogen pentoxide

$$N_2O_5 \rightarrow NO_2 + NO_3$$

The rate constant for this reaction is 0.0314 s^{-1} at 25°C. What is the half-life of N_2O_5? How long would it take for the concentration of N_2O_5 to be reduced by a factor of 5?

14. A graduate student was asked to determine the association (if any) between two methods for the measurement of PCBs in human fat. She obtained the following results (in ppm) for seven different samples using two different analytical methods. Is there a statistical association between the methods, and if so, how strong is it? Excel could be quite handy here.

Sample number	1	2	3	4	5	6	7
Method A	9.0	18.2	17.5	14.2	11.0	10.1	12.2
Method B	7.5	15.5	14.3	12.2	19.0	8.5	9.8

15. During his travels, Professor Hites was on a very fast train (a "TGV") traveling from Paris to Bordeaux, a distance of 420 km; the total train travel time was 2 h and 15 min. While looking out the window, he noticed another TGV train pass, going in the opposite direction, an event which took 3.2 s. How long was the other train, and what was its conductor's given name?

16. An environmental science student from a less than ideal school reported an atmospheric concentration of SO_2 of 75 ppb on a weight per unit weight basis. Please correct this number to the correct units.

17. The concentration of tetrachloroyuckene in a previously unpolluted lake was measured as a function

of time (see the graph below). What is the constant describing the rate of increase of this compound in this lake?

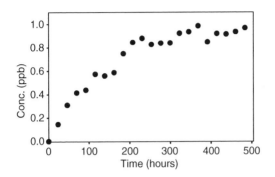

ANSWERS TO THE PROBLEM SETS

PROBLEM SET 1

1. 1.78×10^{-8} cm $= 1.78$ Å
2. 18 ppb
3. ~1 kg or ~2 lb
4. 103 µg/m^3
5. 5.2×10^{18} cm^{-3}, $10^{17.0}$ cm^{-3}
6. 3.5 g
7. 3.4×10^4 t
8. 2.1 t/day
9. 0.14 t
10. 100 g (assumed size of garage was 6500 ft^3)
11. 3.8×10^6 cm^{-3}
12. No! (should be about 2000 trees)
13. 100 years
14. 3.2×10^7 t from tires; 3.0×10^{12} t is the current load

PROBLEM SET 2

1. 10^6 tires; 1 in 200
2. 3×10^8 molecules
3. 80 kg; 2.1 years
4. 0.48 ng/L

Elements of Environmental Chemistry, by Ronald A. Hites
Copyright © 2007 John Wiley & Sons, Inc.

5. 160 mg/year, assuming a house with an area of 2000 ft^2
6. ~15 years
7. ~0.5 ppm, assuming fish weigh 1 kg and bears weigh 1000 kg
8. 2.1 h
9. (b) 91%
10. 680 m^3
11. April 1, 2000
12. 0.60 year
13. 0.18 year
14. 89%
15. 8.4 years; 0.88 ppt
16. 0.20 year
17. 0.19 year^{-1}
18. 1.7 years
19. 87 ng/L and 63 ng/L
20. 11 days

PROBLEM SET 3

1. 266, 40, 480 kJ/mol
2. +0.79 K
3. 290.73 K
4. −3.88 K
5. No (calculated concentration = 50 µg/L)
6. 284.9 K
7. +0.0047
8. 2.2 × 10^{11} kg/year
9. $hc/5k$ = 2883 µm deg

10. 2200 Tg
12. $C_0 = 2.5\%$, $t_{1/2} = 1.5$ h
13. $0.0017 \, \text{h}^{-1} \, \text{mM}^{-1}$
14. 20.6 µg/L, 57 days
15. $4 \times 10^9 \, \text{cm}^{-3}$
16. $10^{4.7} \, \text{cm}^{-3} \, \text{s}^{-1}$; $10^{5.8} \, \text{cm}^{-3} \, \text{s}^{-1}$; 2.7 days
17. 3400 ft, assuming 20% of the hunters are in the field at once
18. 84°C

PROBLEM SET 4

1. 7.57, 8.34, 10.33
2. 3.91
3. 2.4×10^{-4} M
4. 7.7 mg/L
5. 8.14
6. 8.17
7. 9.8×10^{-7} M, 7.6×10^{-5} M
8. $\Delta = -0.070$ pH units
9. 58 ng
10. 180 ng/L
11. 377 mg/L
12. 98 ppm
13. 1.1 Tg/year
14. $0.39 \, \text{h}^{-1}$; let dogs out
15. 2.1 ppb, 0.83 ppb
16. 0.74 cm
17. 1.7×10^7 molecules/cm^3
18. 7.4 mg/L
19. Yes (an additional 1.1 mg/L of O_2 is needed)

PROBLEM SET 5

1. 1.2 ppm
2. 4.23; 3.62; 3.25; 3.01; 2.64; 2.27
3. -0.015 ng m^{-2} year^{-1}
4. 9.6×10^{-4} cm/s; 6.3 m/s
5. 0.72, 1.85 cm/h, 740 ng m^{-2} h^{-1}, 97days;
   600 ng m^{-2} h^{-1}
6. 3×10^{10} m^3/year; 16 ng cm^{-2} year^{-1}
7. 0.0022
8. 4.8 ng cm^{-2} year^{-1}
9. 48 ng/g
10. 26 times higher for the pentachlorobiphenyl
11. 6×10^5
12. 2.2 ng/L
13. 1.8 ng cm^{-2} year^{-1}
14. 14 μg
15. 0.39 t/year
16. 190 ppm; 2.9 days; 40 ppm
17. 238 kg/year; 2.6 ng/L; 4.91
18. 2.4 years

PROBLEM SET 6

1. DDT, Atrazine, Heptachlor, Kepone, Methoprene,
   Treflan, Mirex, Lindane, Methoxychlor, 245-T,
   Dursban, PCP, Chlordane, Glyphosate, Malathion,
   Permethrin
2. 290 ppm; 1.8 h; 69 ppm
3. 3.9 g
4. 2.7 kg; 30 mg
5. Yes

6. 31 days
7. 21 ppb; 15 days; 69 days
8. 17 μmol/L
9. 5.55; 2400 ppm
10. 48 times more lipophilic
12. 1.64×10^7 cm^3/s; 104 s
13. 51 s
14. $r^2 = 0.991$ without outlier
15. ~300 m
16. 34 ppb
17. $(2.5 \pm 0.5) \times 10^{-6}$ s^{-1}

INDEX

Elements of Environmental Chemistry, by Ronald A. Hites
Copyright © 2007 John Wiley & Sons, Inc.